# Industrial Electrostatics:
# Fundamentals and Measurements

# ELECTRONIC & ELECTRICAL ENGINEERING RESEARCH STUDIES

## ELECTROSTATICS AND ELECTROSTATIC APPLICATIONS SERIES

*Series Editor:* **Professor J. F. Hughes,** *Department of Electrical Engineering, University of Southampton, England*

**Other titles in the Series include:**

4. Electrostatic Powder Coating
   **J. F. Hughes**

6. Computer Modelling in Electrostatics
   **D. McAllister, J. R. Smith** *and* **N. J. Diserens**

8. Electrostatic Damage in Electronics: Devices and Systems
   **William D. Greason**

9. Electrostatic Hazards in Powder Handling
   **Martin Glor**

10. Electrostatic Spraying of Liquids
    **Adrian G. Bailey**

11. Computation of Lightning Protection
    **Tibor Horváth**

12. Electrostatic Discharge in Electronics
    **William D. Greason**

13. Industrial Electrostatics: Fundamentals and Measurements
    **D. M. Taylor** *and* **P. E. Secker**

**U H** University of Hertfordshire

Learning and Information Services
Hatfield Campus Learning Resources Centre
College Lane Hatfield Herts AL10 9AB
Renewals: Tel 01707 284673 Mon-Fri 12 noon-8pm only

This book is in heavy demand and is due back strictly by the last date stamped below. A fine will be charged for the late return of items.

ONE WEEK LOAN

# Industrial Electrostatics:
# Fundamentals
# and Measurements

## D. M. Taylor
*School of Electronic Engineering and Computer Systems*
*University of Wales, Bangor, UK*

*and*

## P. E. Secker
*Electronic Engineering Group, Department of Physics*
*Keele University, UK*

**RESEARCH STUDIES PRESS LTD.**
Taunton, Somerset, England

**JOHN WILEY & SONS INC.**
New York · Chichester · Toronto · Brisbane · Singapore

# RESEARCH STUDIES PRESS LTD.
24 Belvedere Road, Taunton, Somerset, England TA1 1HD

**Marketing and Distribution:**

*Australia and New Zealand:*
Jacaranda Wiley Ltd.
GPO Box 859, Brisbane, Queensland 4001, Australia

*Canada:*
JOHN WILEY & SONS CANADA LIMITED
22 Worcester Road, Rexdale, Ontario, Canada

*Europe, Africa, Middle East and Japan:*
JOHN WILEY & SONS LIMITED
Baffins Lane, Chichester, West Sussex, England

*North and South America:*
JOHN WILEY & SONS INC.
605 Third Avenue, New York, NY 10158, USA

*South East Asia:*
JOHN WILEY & SONS (SEA) PTE LTD.
37 Jalan Pemimpin 05-04
Block B Union Industrial Building, Singapore 2057

**Library of Congress Cataloging-in-Publication Data**

Taylor, D. M. (David M.)
    Industrial electrostatics : fundamentals and measurements / D.M.
Taylor and P.E. Secker.
        p.   cm. — (Electronic & electrical engineering research
studies. Electrostatics and electrostatic applications series ; 13)
    Includes bibliographical references and index.
    ISBN 0-471-95233-8 (John Wiley & Sons). — ISBN 0-86380-158-7
(Research Studies Press Ltd.)
    1. Industrial electronics.  2. Electrostatics.  I. Secker, P. E.
(Philip Edward), 1936–   .  II. Title.  III. Series.
TK7881.T39   1994
537'.2'02462—dc20                                94-16519
                                                     CIP

**British Library Cataloguing in Publication Data**

A catalogue record for this book
is available from the British Library.


ISBN 0 86380 158 7 (Research Studies Press Ltd.)
ISBN 0 471 95233 8 (John Wiley & Sons Inc.)

Printed in Great Britain by SRP Ltd., Exeter

# Contents

**Editorial Foreword**         **xi**

**1 INTRODUCTION**         **1**
1.1 Industrial Problems Caused by Static Electricity 1
1.2 Applications of Static Electricity 4
1.3 Insulators 12
1.4 Standards 12
1.5 Concluding Remarks 14

**2 BASIC ELECTROSTATIC THEORY**         **16**
2.1 Electrostatic Charges 17
2.2 Forces Between Charges 17
    *2.2.1 Coulomb's Law 18*
2.3 Electric Field 19
    *2.3.1 Electric Field of a Point Charge 20*
    *2.3.2 Flux Density (Displacement) 20*
    *2.3.3 Electric Field of an Infinite Sheet of Charge 21*
    *2.3.4 Electric Field of a Volume Charge 23*
        *2.3.4.1 Gauss' Theorem 24*
        *2.3.4.2 Electric Fields in Spherical Containers 25*
        *2.3.4.3 Electric Field in a Pipe 26*
2.4 Potential 28
    *2.4.1 Potential of a Point Charge 28*
    *2.4.2 Potential of an Infinite Sheet of Charge 29*
    *2.4.3 Equipotentials 30*
2.5 Fields and Potentials Associated with a Volume Charge 30
2.6 Boundaries of Dielectrics 33
    *2.6.1 General Rules at Dielectric Interfaces 34*
        *(a) Electric Fields Normal to the Interface 34*
        *(b) Electric Fields Parallel to an Interface 35*
        *(c) Interfacial Charges 36*
    *2.6.2 Image Charges 36*
        *(a) Image Charges in Conducting Surfaces 36*

v

*(b) Image Charges in Dielectric Surfaces 38*
*2.6.3 Method of Images 40*
   *(a) Line Charge Parallel to an Earthed Plane 40*
   *(b) Sheet of Charge Parallel to an Earthed Plane 41*
   *(c) Image of a Charged Sheet Tangential to an Earthed Roller 42*
2.7 Capacitance 43
  *2.7.1 Parallel-Plate Capacitor 44*
  *2.7.2 Cylindrical and Spherical Capacitors 45*
   *(a) Coaxial Cylinders 46*
   *(b) Parallel Cylinders 47*
   *(c) Cylinder and a Plane 48*
   *(d) Concentric Spheres 49*
   *(e) Sphere and a Plane 50*
  *2.7.3 Other Types of Capacitances 52*
  *2.7.4 Addition of Capacitances 52*
   *(a) Parallel Combination of Capacitors 52*
   *(b) Series Combination of Capacitors 53*
  *2.7.5 Energy Stored in a Capacitor 53*
2.8 Independent and Dependent Electrostatic Parameters 55
2.9 Insulating Materials 57
  *2.9.1 Relative Permittivity (Dielectric Constant) 57*
  *2.9.2 Volume Resistivity 59*
  *2.9.3 Surface Resistivity 60*
  *2.9.4 Relaxation Time 60*
  *2.9.5 Transparency to Electric Flux 63*
  *2.9.6 Breakdown Strength 64*
2.10 Equivalent Circuit of an Electrostatic System 64

**3 INITIATION OF ELECTROSTATIC PHENOMENA   67**
3.1 Charging of Liquids in Pipeline Flow 68
  *3.1.1 The Double-Layer 68*
  *3.1.2 The Streaming Current 73*
   *3.1.2.1 Streaming Current in Long Pipes 74*
   *3.1.2.2 Effect of Flow Velocity 76*
   *3.1.2.3 Effect of Pipe Length 77*
   *3.1.2.4 Effect of Liquid Conductivity 78*
   *3.1.2.5 Outlet Effects 79*
  *3.1.3 Streaming Current in Transformers 80*
  *3.1.4 Electrification of Polymers During Extrusion 81*

*3.1.5 Streaming Current in Insulating Pipes 81*
*3.1.6 Other Related Electrification Mechanisms in Liquids 82*
3.2 Contact and Frictional Charging of Solids 82
    *3.2.1 Models Based on Electron Transfer 85*
        *3.2.1.1 Metal-Metal Contact 85*
        *3.2.1.2 Metal-Insulator Contact 87*
            (a) Surface States 91
            (b) Nature of Surface States 93
        *3.2.1.3 Insulator-Insulator Contacts 95*
    *3.2.2 Triboelectrification 96*
    *3.2.3 Ion Transfer 96*
    *3.2.4 Role of Back-Discharges and Back-Tunnelling*
        *in Tribocharging 98*
    *3.2.5 Tribocharging Limits in Industrial Processes 100*
        (a) Insulating Sheet 100
        (b) Spherical Particles 102
    *3.2.6 Tribocharging in Industry 103*
        (a) Applications 103
        (b) Hazards and Problems 103
3.3 Charging by Induction 104
    *3.3.1 Ink-Jet Printers 106*
    *3.3.2 Atomisation 107*
    *3.3.3 Induction Charging of Personnel 109*
3.4 Corona Charging 110
    *3.4.1 The Corona Discharge 110*
        *3.4.1.1 Corona Threshold Voltage 112*
        *3.4.1.2 Current-Voltage Characteristic of a Corona 114*
        *3.4.1.3 Current Density of a Corona 115*
    *3.4.2 Corona Charging of Airborne Particles 116*
        *3.4.2.1 Limiting Charge 116*
        *3.4.2.2 Charging Time Constant 118*
        *3.4.2.3 Bipolar Charging 120*
        *3.4.2.4 Diffusion Charging 121*
        *3.4.2.5 Back-Ionisation 121*
    *3.4.3 Charging of Insulating Sheets 122*
    *3.4.4 Static Eliminators 124*
    *3.4.5 Nature of Corona Ions 124*
3.5 Other Sources of Static 125

**4 MEASUREMENT OF ELECTRIC FIELD**                    **129**
   4.1   Introduction 129
      *4.1.1  Measurement Philosophy 130*
   4.2   The Induction Probe 130
      *4.2.1  Effects of Atmospheric Currents on Field*
         *Measurement 135*
   4.3   The Field Mill 136
      *4.3.1  Detecting Electric Field Polarity 139*
      *4.3.2  Field Penetration through Apertures 143*
      *4.3.3  Making Field Mill Measurements 144*
      *4.3.4  Field Mill Applications 148*
   4.4   Vibrating  Probe Field Meter 152
   4.5   Other Field Detection Techniques 153
      *4.5.1  The Ballistic Probe 153*
      *4.5.2  The Charged Particle Probe 155*
      *4.5.3  Electro-Optical Effects 155*

**5 MEASUREMENT OF VOLTAGE**                    **159**
   5.1   Introduction 159
   5.2   Voltmeters Exploiting Electrostatic Force Effects 159
   5.3   Field Mill Voltmeter 167
   5.4   Voltage Dividers 170
      *5.4.1  Direct Voltage Dividers 171*
      *5.4.2  Alternating Voltage Dividers 175*
      *5.4.3  Dividers for Impulse and Step-Function Voltages 177*
   5.5   Spark Gaps 179
   5.6   Other Techniques for Voltage Measurement 182

**6 MEASUREMENT OF CHARGE**                    **185**
   6.1   Charged Particles 185
   6.2   Charged Surfaces 193
      *6.2.1 Achieving Good Spatial Resolution in Surface*
         *Charge Measurements 196*
   6.3   Charge Distributed Throughout a Volume 201
      *6.3.1  Space-Charge Measurements in a Gaseous Medium 203*
      *6.3.2  Space-Charge Measurement in Liquid 204*
      *6.3.3  Volume Charge Density in Solid Materials 205*
   6.4   Concluding  Comments 207

**7 RESISTANCE AND CHARGE DECAY**      **210**

7.1   Introduction 210

7.2   Bonding to Ground 211

    *7.2.1   Practical Monitoring of Earth Bonding 211*

       *7.2.1.1 Antistatic Boots 211*

       *7.2.1.2 Buried Conductors 212*

       *7.2.1.3 Work-Pieces on Conveyors 214*

7.3   Characterisation of Process Materials 215

    *7.3.1   Volume Resistivity 215*

       *7.3.1.1 Volume Resistivity of Solids 216*

       *7.3.1.2 Volume Resistivity of Liquids 217*

       *7.3.1.3 Volume Resistivity of Powders 218*

       *7.3.1.4 General Considerations 220*

    *7.3.2   Surface Resistivity 221*

       *7.3.2.1 General Considerations 222*

    *7.3.3   Charge Decay Measurements 223*

       *7.3.3.1 Liquids 223*

       *7.3.3.2 Films and Laminates 224*

       *7.3.3.3 On-Line Surface-Charge-Decay Gauge 228*

       *7.3.3.4 General Considerations 230*

**8 ELECTROSTATIC DISCHARGES (ESD)**      **232**

8.1   Electrical Breakdown in Air 232

    *8.1.1   Townsend α-Process 232*

    *8.1.2   Townsend γ-Process 233*

8.2   Types of Electrical Discharge 234

    *8.2.1   Spark Discharge 235*

    *8.2.2   Brush Discharge 236*

    *8.2.3   Corona Discharge 238*

    *8.2.4   Propagating Brush Discharge 238*

    *8.2.5   Cone Discharges 240*

8.3   Simulating Spark Discharges 242

    *8.3.1   Minimum Ignition Energy 242*

       *8.3.1.1 Incendivity of Electrostatic Discharges 242*

         (a) Spark Discharge 242

         (b) Brush Discharge 243

         (c) Corona Discharge 244

         (d) Propagating Brush Discharge 245

         (e) Cone Discharge 245

       *8.3.1.2 Measuring the MIE of a Gas/Air Mixture 245*

*8.3.1.3 Measuring the MIE of a Dust/Air Mixture 247*
    *8.3.2  Human Body Model 250*
8.4  Monitoring Spark Discharges 253
    *8.4.1  Radio Frequency Detection 253*
    *8.4.2  Measuring the Current Flow in*
        *Electrostatic Discharges 256*
8.5  Concluding Remarks 261

**Author Index**                                                264

**Subject Index**                                               267

# *Editorial Foreword*

The previous twelve titles in this series have described various aspects of electrostatics, all of which, in one way or another, have relied on the use and understanding of a number of electrostatic phenomena and different measurement techniques. In this monograph, the origin of electrostatic phenomena as well as fundamental parameters such as resistivity, electric field, potential, charge etc. are discussed and techniques for measuring these parameters described in detail. Throughout an emphasis is placed on relating each new topic to industrial situations.

The authors not only describe methods of measurement but also clearly identify potential errors in otherwise seemingly simple techniques. This, combined with useful discussions relating to the interpretation of data, makes this monograph not only a useful companion to other titles in the series but also an extremely valuable reference in its own right.

Whether for teaching, for research or for industrial applications, following the guidance given by the authors will ensure that any measurement made on an electrostatic system will be a meaningful one.

**J.F.Hughes**
**S.C.Johnson Professor of New Technologies**
**Southampton, March 1994**

*CHAPTER ONE*

# INTRODUCTION

For many scientists and engineers Static Electricity conjures up images of ancient Wimshurst machines, Van de Graaff generators and "hair-raising" experiments seen during school visits to a science museum. Yet the phenomenon is much more common than this rather narrow view might indicate. Everyone at some time or another has experienced, and will continue to experience, the effects of static electricity at first hand. Who, for example, has not been fascinated by the awesome power of lightning storms? And who has not seen or heard the crackling of electrical discharges when undressing on cold, dry winter days? In our well-appointed world, electric shocks from door knobs in hotels or from filing cabinets in offices furnished with plush, highly insulating carpets are commonplace, as are shocks from motor cars. The electrostatic attraction of dust to plastic materials is a regular occurrence and careful scrutiny of the surface of plastic goods left on shelves or in storage for any length of time will reveal beautiful, finely feathered Lichtenberg patterns; though the phenomenon goes unappreciated by manufacturers of food containers and "white" goods!

## 1.1 INDUSTRIAL PROBLEMS CAUSED BY STATIC ELECTRICITY

The everyday manifestations of static electricity mentioned above are minor irritants compared with the problems caused by the same effects in industry. Severe production losses can occur when, as a result of electrostatic cling, plastic films or textile fibres wrap themselves around rollers or hang up on product guides causing line shutdown. During high speed film winding operations, bright, luminous sparks are often seen propagating along the surfaces of plastics and paper products. These are caused when the magnitude of the static charge on the product is so great that the air in the vicinity of the product fails electrically. The occurrence of such sparks would obviously lead to problems where photographic or x-ray film is being processed if no precautions were taken to limit static build-up. Manufacturers of plastic film capacitors and magnetic tape must also take steps to limit static generation since degradation of the product may easily occur, either directly from spark damage or indirectly from the attraction of dust to the product. Instances of sparks from film wind-up reels and from bins of insulating powder products causing shock and even severe injury to personnel are widespread.

The effects of such events are compounded if they occur in or near a flammable or explosive atmosphere where the risk of electrostatically-induced fires and explosions is ever present. It is essential, therefore, that the petrochemical, plastics and pharmaceutical industries adopt effective static protection measures that minimise and hopefully eliminate altogether the risk of electrostatically-induced ignitions in the manufacture, handling and transport of their products.

Whether or not electrostatic sparks can cause ignition depends on a number of factors including the minimum ignition energy (MIE) of the flammable vapour or powder cloud, the type and the source of the discharge as well as the energy of the spark and the rate at which this energy is dissipated.

Because of the complexity of the physics and chemistry of ignition and indeed the continuing debate concerning the most appropriate method for measuring the MIE, predicting the conditions under which explosions may occur becomes a matter of expert judgement. Nevertheless, since isolated metal conductors and ungrounded personnel are the most common sources of electrostatic sparks, simple measures such as earth bonding reduce significantly any explosion risk.

The scaling up of industrial processes requires special consideration to be given to the electrostatic consequences. For example, the electric fields accompanying electrostatically charged powder or droplet clouds increase as the volume of the containing vessel increases (section 2.3.4.2), thus enhancing the risk of an electric-al discharge in the container. This was a contributory factor (Van der Meer, 1971) in the explosions that ripped open three very large crude carriers (VLCC) in December 1969, damage that was sufficient to sink one of them. The high potentials and electric fields which led to the incidents were a consequence of the highly charged mist created during the washing of cargo holds with high-pressure water jets.

Increasing process speed also exacerbates most electrostatic problems. This is a consequence of increasing the charge generation rate to the point where it exceeds the rate at which it can dissipate safely to ground. Thus, processes that had hitherto been free of electrostatic problems can become susceptible as the process is upgraded. Furthermore, since the natural dissipation rate of static is sensitive to ambient atmospheric conditions, certain industrial processes are particularly vulnerable to changes in relative humidity, with electrostatic problems appearing to come and go in a random and inexplicable manner.

Often, static problems in industry arise from minor and seemingly unimportant changes to the processing equipment. For example, the insertion of short sections of plastic pipe into otherwise all metal pipework (and vice versa) can lead to the presence of non-grounded metal assemblies which readily charge to the sparking potential of the surrounding air. In film winding operations, the replacement of metal or conducting rubber rollers with insulating rubber or ceramic rollers can, unwittingly, lead to a significant increase in the rate at which static is generated. Frequently, it is the incorrect location of static neutralisers that is the cause of significant electrostatic charging. Therefore, it is particularly important that when changes are made, whether these are temporary or permanent, to an electrostatically sensitive process the modifications introduced should be properly audited to avoid the introduction of new electrostatic problems.

In the last decade or so, as a consequence of rapid improvements in integrated-circuit processing technologies, the semiconductor industry has become the latest in a growing number of modern industries to suffer immense production losses from the effects of static electricity. The increasing use of furnishings, packaging and clothing manufactured from plastics and man-made fibres has resulted in the creation of an environment in which electrostatic charges are readily generated. The small line-widths (1μm and less) typically employed in today's integrated circuits means that electronic devices are particularly vulnerable to the effects of electrostatic discharges (ESD).

The most common cause of device failure is an electrical discharge (spark) from personnel directly to the pins of the device. The current from the spark can lead to such high current densities in the device that, for example, interconnects evaporate and junctions fail. The high voltages associated with the spark can cause the electrical breakdown of the thin oxide insulation present under the gate electrode of devices based on MOS technology. These oxides are typically 100 nm thick and since the breakdown strength of the oxide is $8\times10^8$ V/m the layer fails when voltages in excess of 80 V appear across them. Ungrounded personnel may easily charge to several thousand volts simply by walking over carpets or even by shuffling around while seated on a PVC covered chair (Unger, 1981).

Limited on-board protection can be provided by means of special circuits connected to each pin (Enoch et al, 1983). However, this is expensive in space on the chip and invariably degrades the high frequency performance of the device: protection measures usually include the introduction of resistances and capacitances at the input with the aim of dissipating the spark energy and reducing the spark current. As processing technology improves and submicron devices become more commonplace, ESD sensitivity will increase further and stringent measures will need to be taken to ensure that electronic devices and systems are manufactured, operated and serviced in an environment which is essentially free of static electricity.

Clearly then, for many industrial situations there is a need to understand the phenomenon of static electricity in order to determine the most appropriate countermeasures for reducing its disruptive effects on industrial processes and particularly to reduce the hazards associated with it.

## 1.2  APPLICATIONS OF STATIC ELECTRICITY

In contrast to the undesirable and the potentially dangerous aspects of static electricity mentioned above is the pivotal role of electrostatic phenomena in a number of extremely successful and highly profitable commercial processes.

Pre-eminent among these is electrophotography as epitomised by the Xerox process invented by Chester Carlson for which a patent was filed in 1939 and granted in 1942.

The various steps in the process are outlined in Figure 1.1. Initially a photo-conductive film,e.g. selenium, is sensitised by exposure to a corona discharge from a corotron (see section 3.4.3) which charges the film uniformly over its surface (Figure 1.1(a)). An optical image of the material to be copied is then focused onto the uniformly charged photoconductor. The conductivity of the illuminated regions of the photoconductor is  increased significantly so that the surface charge in these regions is conducted away to the earthed substrate as in Figure 1.1(b). The charge pattern remaining on the photoconductor after exposure is known as the latent image and is developed by dusting the surface with a mixture of toner and carrier particles. When the carrier particles rub against the smaller toner particles, the latter acquire an electrostatic charge of opposite sign to that used for charging the photoconductor. Therefore, during the dusting step, toner particles adhere to charged areas on the photoconductor surface (Figure 1.1(c)), thereby developing the latent image. In the final stages of the process, the developed image is transferred to paper, a process aided by exposing the rear surface of the paper to a further corona discharge as shown in Figure 1.1(d). Finally, the developed image is fixed by heating (Figure 1.1(e)). Full details of the process and the background technology can be found in the texts by Dessauer and Clark (1965) and Schaffert (1975).

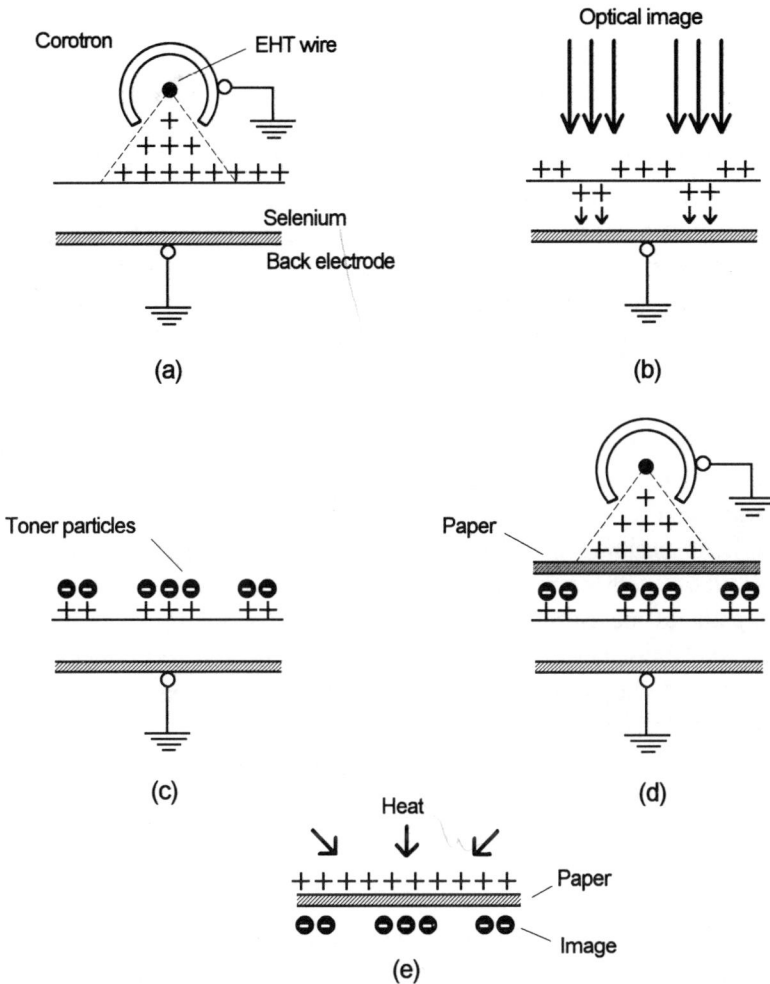

**Figure 1.1** *Stages in the production of an electrophotographic image. (a) Sensitisation of the photoreceptor (b) exposure to the optical image (c) development of the latent image (d) transfer of developed image to paper and (e) fixing of image.*

The principles of electrophotography outlined above have now been extended to colour photocopiers and more recently to a new generation of "quiet" printers. These laser printers, in which the digitised image to be printed is "written" on the photoconductor by a scanning laser beam or a laser diode array, are rapidly replacing the noisy mechanical printers hitherto associated with computers and word-processors.

A parallel development is ink-jet printing (section 3.3.1) which provides a simple method of applying alphanumeric and bar codes to a range of household goods (Sweet, 1965). This technique is also being extended into the "quiet" computer printer market and is capable also of good quality colour reproduction. In these printers a train of ink droplets is created at a fine nozzle typically using ultrasonic vibration. At the nozzle exit, just before they break away, the droplets are exposed to an electric field and become charged by induction. "Writing" is accomplished by electrostatic deflection of the charged droplets when they pass between a pair of plates to which appropriate voltages are applied.

A related process is electrostatic paint spraying in which drops of liquid paint are charged to such a high level that the electrostatic repulsion forces in the droplet overcome surface tension forces,causing atomisation. The finely charged mist thus created drifts to the grounded workpiece in the electric field produced by the high voltage spray gun.

Electrostatic charging of particles is the first step in dry powder coating (Hughes, 1985), another important application of electrostatics. Here, the powder or paint particles are electrically charged by triboelectrification or by passing through a corona discharge in a spray gun. The charged particles are then directed by the combined effects of gas flow and electric field to an earthed workpiece. When the particles approach the workpiece, electrical forces overcome the aero-dynamic forces so that powder precipitates onto the workpiece and is held there by image forces. The powder coat is then cured by heating. Dry powder coating using this technique has several advantages over conventional methods, the main ones being

(i) a uniform coating can be achieved because mutual repulsion of similarly charged powder particles reduces the tendency for particles to accumulate at one point,

(ii) electrostatic precipitation of the powder encourages "wrap-around",i.e. surfaces not in direct line-of-flight also become coated,although judicious use of aerodynamics is necessary to obtain the best results when painting recesses, and

(iii) little wastage occurs since excess powder can be collected, using the electric curtain method, for example, and recycled.

The electric curtain is a system developed by Masuda and co-workers in Tokyo (Masuda et al, 1972; Masuda and Matsumoto, 1974) for suspending and transporting charged particles. In its simplest form the curtain consists of a parallel array of cylindrical electrodes, insulated from each other and connected to a source of 3-phase alternating voltage as shown in Figure 1.2. The electric field above the electrode array is in the form of a travelling wave which can levitate and transport charged particles from left to right above the surface of the array.

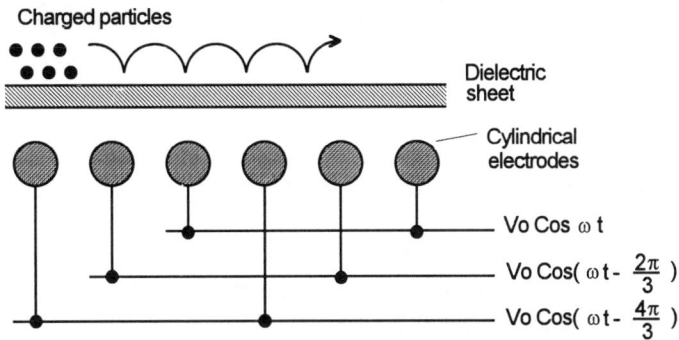

**Figure 1.2**     *Masuda's 3-phase electric curtain for transporting charged particles.*

A very similar process to electrostatic powder coating is electrostatic precipitation. Applications range from the removal of fly ash from the flue gases of coal fired power stations to the removal of cigarette smoke from rooms. Electrostatic precipitation is particularly advantageous for particles in the size range from about 100 μm down to 1 μm which quickly block fabric filters causing a large pressure drop across them as the particles are collected.

In conventional electrostatic precipitators, particles are charged during their passage through a corona discharge as in Figure 1.3. The charged particles are then subjected to an electric field which directs them to the walls of the flue or ducting where they precipitate out of the gas flow. Since the need for a filter has been obviated in such systems, gas flow rate is high and remains high during particle collection. To maintain efficient performance, the precipitated particles must be removed periodically. Various arrangements have been commercialised, including single and two-stage systems, and novel charging methods such as the boxer charger have been developed. Interestingly, while conventional precipitators employ a DC corona to achieve unipolar charging, the boxer charger developed by Masuda and co-workers (Masuda et al, 1978) achieves the same result using an AC technique.

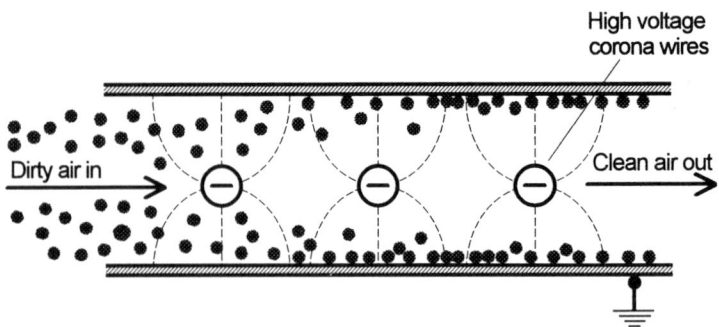

**Figure 1.3**    *Section through a conventional electrostatic precipitator. The particles are corona charged and drift in the electric field to the walls of the precipitator.*

The basis of the boxer charger (Figure 1.4) is the surface discharge (or plasma) which is initiated at the fingers of a digitated electrode by a high frequency (10 kHz) alternating voltage applied between the fingers and a back plate. Two sets of

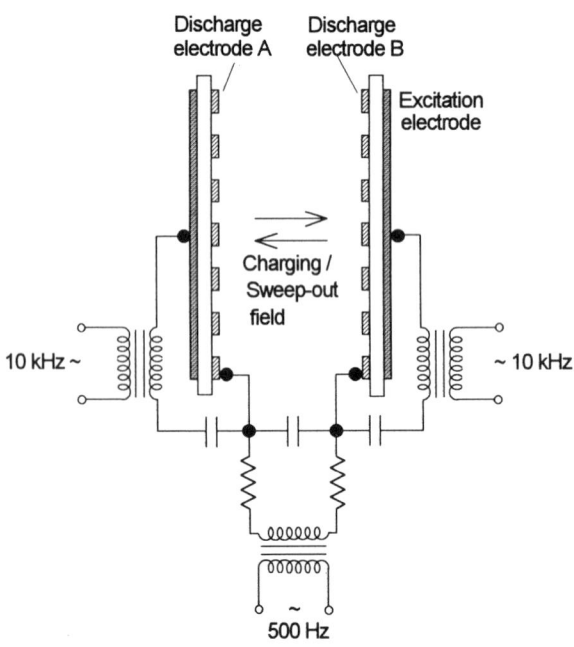

**Figure 1.4**    *Masuda's boxer charger concept.*

such electrodes are arranged to face each other. A low-frequency charging voltage (at say 500 Hz) is applied across the air-gap between the two electrode sets. This low-frequency voltage serves to both charge the dust particles and sweep them out of the air flow.

Using the circuit arrangement shown, short bursts of the high-frequency excitation voltage (say 10 kHz) are applied alternately to each pair of discharge electrodes. By suitably adjusting the phase of these high-frequency bursts relative to the low-frequency charging voltage it can be arranged that positive ions are emitted sequentially from the surface plasmas induced at the two digitated electrodes on alternate half cycles of the charging voltage. (The relevant timing diagrams are shown in Figure 1.5). In this way, any particles flowing through the charger in one half cycle will be bombarded by positive ions, say, from the left hand electrode and in the second half cycle by positive ions from the right hand electrode. Masuda coined the term boxer charger to describe this "two-handed bombardment!

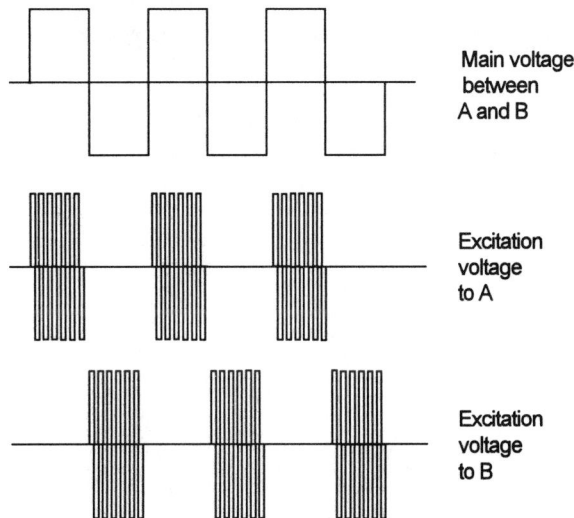

**Figure 1.5**    *Timing diagram for the alternating voltages applied to the various electrodes in Masuda's boxer charger.*

The cycling of the main field ensures that very little precipitation actually occurs in the charger. Consequently, because the charger can operate at higher current densities and higher fields, greater operating efficiencies are possible than with conventional corona precipitators. Furthermore, by simply changing the phase of the excitation signals to the discharge electrodes negative charging can be achieved.

Electrostatic precipitators have been used successfully for most of this century and, despite their high efficiency, significant research effort is still continuing to further improve their performance. A major goal is to enhance the collection efficiency for sub-micron particles as well as the agglomerated oxides of sulphur and nitrogen which are significant contributors to acid rain.

Electrostatics is used to good effect in particle separation and is especially useful in mineral beneficiation. Particles may be separated on the basis of their size, shape, electrical resistivity and even the sign of the tribocharge generated on their surfaces. Again, the first step in the process is to charge the particles triboelectrically, or perhaps by corona or induction. Separation of the particles then takes place in an electric field which acts in the manner of a mass-spectrometer, separating the particles on the basis of their charge-to-mass ratio.

A very simple electrostatic separator is shown in outline in Figure 1.6. Finely divided particles are slowly poured from a hopper onto an earthed conducting drum. All the particles are immediately corona-charged during their passage under

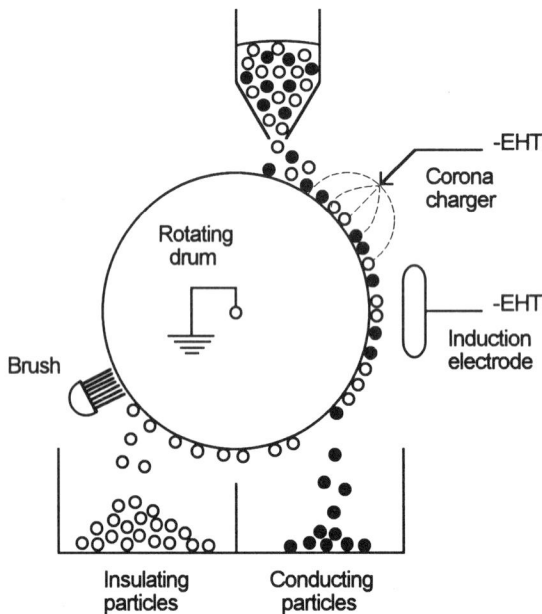

**Figure 1.6**    *A simple electrostatic separator which uses corona and induction charging to effect separation of insulating and conducting particles.*

a sharply pointed high-voltage electrode. The conducting particles quickly lose their charge resulting in a reduction in the image force holding them to the drum. Consequently, as the drum rotates these particles fall off into collecting bins located under the drum. This process may be helped by incorporating an induction electrode to pull off the conducting particles by reverse charging. The more insulating particles remain attached to the drum until they are removed by brushing.

The principles of operation of a number of other different types of electrostatic separator are described by Cross (1987).

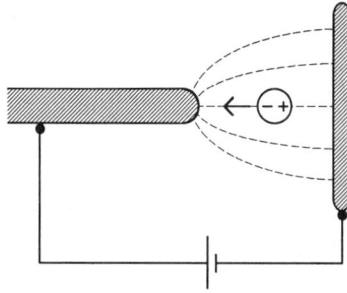

**Figure 1.7**    *Migration of an electrically neutral particle in a nonuniform electric field*

Electrostatic processes, such as dielectrophoresis, are beginning to impact the expanding biotechnology industry where the control and separation of biological cells is of great importance. Dielectrophoresis is a process which relies on the fact that an electric field induces a polarisation within a particle, in this case a cell. If the polarised particle is located in a non-uniform electric field, Figure 1.7, it will migrate towards the high field region. The degree of polarisation and indeed its frequency-dependence are functions of the "state" of the cell, so that the technique can be used for concentrating biological cells of one type even when they are mixed with high concentrations of other types of cell. The separation of live cells from dead cells by this method was demonstrated some years ago by Crane and Pohl (1968). The remarkable progress made since in developing this technology can be seen in Figure 1.8 (a) which shows yeast cells collecting in the high field zones of an interdigitated electrode array (Pethig et al, 1992).By applying an AC voltage of different frequency the cells can be induced to migrate to the low field regions (b).

It should be clear even from the few illustrative examples above that electro-statics plays an important role in a range of successful commercial ventures. The range of applications described is by no means exhaustive. For example, no mention has been made of electrostatic fluidised beds, electrostatic containment and control of particles and aerosols, electrostatic generators, ferroelectricity, pyroelectric detectors, piezoelectric sensors and many other applications. For a more comprehensive review, the reader is referred to the texts by Cross (1987) and Moore (1973).

## 1.3 INSULATORS

A common link throughout industrial electrostastics is the nature of the material being processed. Whether static electricity gives rise to a problem or is used as an integral part of the commercial process, at some stage during manufacture, handling or storage, an electrically insulating material,e.g. plastic film, hydrocarbon oil, organic powder or any material insulated from earth,is being electrically charged.

In liquids, the carriers of electrical charges are generally ionic in nature while in solids it is usually electrons (and perhaps even holes) that are the important charge carriers. In gases, both electrons and ions are important in the charge transport process. The poor electrical conductivity of insulating materials arises from a paucity of free charge carriers in them. This is because insulators are materials which are very stable electrically. They are not easily ionised to yield free electrons,and impurities contained in them do not readily dissociate into positive and negative ions. When electrons or ions are present in insulating liquids, the speed with which they can migrate through the material is generally low, the mobility of ions being limited by their inherent size and the viscosity of the medium. In solids, electrons become localised at defects known as traps.

It is the inability of insulating materials to allow charge migration that leads to the accumulation of unwanted static charge in manufacturing processes and, perversely, is also the property necessary for the efficient operation of most electrostatic-based technologies.

## 1.4 STANDARDS

The impact of electrostatics on industry is so great that a large number of National and International Standards have been issued over many years. Several of these are concerned with ensuring safe operation in the presence of flammable and explosive materials. Standards have also been issued for the handling of static sensitive electronic devices as well as for defining and measuring the relevant electrical properties of insulating, static-dissipative and conductive materials for use as furnishing materials, flooring, footwear and clothing.

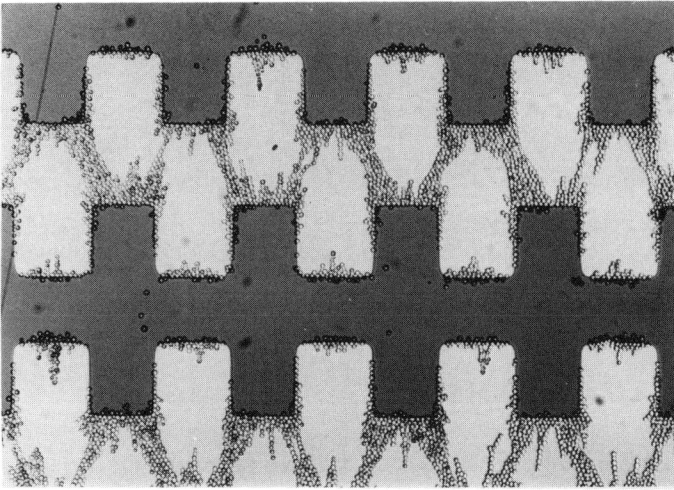

(a)

(b)

**Figure 1.8**     *Micrographs showing the existence of (a) positive dielectrophoresis where particles (in this case yeast cells) collect in the high field region and (b) negative dielectrophoresis where they collect in the low field regions. (Courtesy of Professor R. Pethig)*

In the UK, a particularly useful standard is BS 5958 (Control of Undesirable Static Electricity, Parts 1 and 2) which provides a review of electrostatic phenomena but,more importantly, provides sound practical advice for dealing with the problems that arise in a number of specific industries. Those in the electronics industry should be aware of BS 5783 (Code of Practice for Handling Electrostatic Sensitive Devices) and DOD-HDBK-263 issued by the US Department of Defense which suggest appropriate methods for dealing with the problems posed by static electricity to sensitive electronic devices. During the writing of this book a more comprehensive code of practice (CENELEC-CECC 00015) covering the problem of ESD in the electronics industry was issued.

## 1.5 CONCLUDING REMARKS

The safe handling of electrostatic-sensitive materials and devices requires a thorough appreciation of electrostatic phenomena. The assessment of risks associated with a particular process must involve accurate, reliable measurements of electrostatic properties such as charge, electric field, potential, material resistivity etc. This is true, also, of any industrial process using electrostatics. Only when the process is thoroughly understood and characterised by measurements can it be optimised so that it operates at peak efficiency.

In the next two chapters, the basic theory of electrostatics is treated in a way which relates directly to industrial processes, with several examples being given of the more common phenomena that are observed. The remaining chapters are devoted to measurement techniques used for quantifying important electrostatic parameters, together with a discussion of the procedures necessary for assuring that the results are accurate and reliable. Thus a solid foundation is provided for analysing electrostatic systems and for solving the more common problems that arise.

## REFERENCES

Crane JS and Pohl HA 1968 "A study of living and dead yeast cells using dielectrophoresis" *J Electrochem Soc* **115** 584-586.

Cross JA 1987 *"Electrostatics: Principles, Problems and Applications"* (Bristol: Adam Hilger).

Dessauer JH and Clark HE eds 1965 *"Xerography and Related Processes"* (London and New York: Focal Press).

Enoch RD, Shaw RN and Taylor RG 1983 "ESD sensitivity of NMOS LSI circuits and their failure characteristics" *Proc EOS/ESD Symposium* **EOS-5** 185-197.

Hughes JF 1985 *"Electrostatic Powder Coating"* (Letchworth: Research Studies Press and New

York: Wiley).

Masuda S and Matsumoto Y 1974 "Contact-type electric curtain for electrodynamical control of charged dust particles" *Dechema Monograph* **72** 293-301.

Masuda S, Fujibayashi K, Ishida K and Inaba H 1972 " Confinemant and transportation of charged aerosol clouds via electric curtain" *Electron Eng Japan* **92** 43-52.

Masuda S, Washizu M, Mizuno A and Akutsu K 1978 "Boxer charger - a novel charging device for high resistivity powders" *Proc IEEE/IAS* 16-22.

Moore AD ed 1973 *"Electrostatics and its Applications"* (New York: Wiley).

Pethig R, Huang Y, Wang X-B and Burt JPH 1992 "Positive and negative dielectrophoretic collection of colloidal particles using interdigitated castellated microelectrodes" *J PhysD:Appl Phys* **25** 881-888.

Schaffert RM 1975 *"Electrophotography"* (London and New York: Focal Press).

Sweet RG 1965 "High frequency recording with electrostatically deflected ink jets" *Rev Sci Instrum* **36** 131-136.

Unger BA 1981 "Electrostatic discharge failures of semiconducting devices" *IEEE/Proc IRPS* 193-199.

Van der Meer D 1971 "Electrostatic charge generation during washing of tanks with water sprays - I: General Introduction" *IOP Conf Ser No 11* 153-157.

## CHAPTER TWO

# BASIC ELECTROSTATIC THEORY

Static electricity is perceived generally as a highly non-reproducible phenomenon. Yet, if this were the case, the successful industries described in Chapter 1 which rely heavily on electrostatics would not have evolved. Surprising though it may seem to those experiencing problems caused by static, it is a subject which follows well-known physical laws, and a working knowledge of even some of the basic ideas can help in overcoming problems and in understanding the more common electrostatic occurrences. In this Chapter then, the aim is to introduce the reader to those theoretical concepts which are fundamental to the understanding of industrial electrostatics. Examples are given also of the practical situations to which each new theoretical concept is relevant.

From these considerations a description of an electrostatic system emerges in terms of a simple electrical equivalent circuit model. This circuit analogy is useful since it allows predictions to be made concerning the behaviour of an electrostatic system and shows clearly why electrostatic phenomena are so dependent on environmental factors such as humidity, cleanliness etc.

Electrostatics is, essentially, the science of interacting electrical charges. In most industrial processes the charges are in motion because they are entrained in or on the product. However, in most cases the system of charges as a whole may be considered as quasi-stationary since its evolution with time is relatively slow. The interactions of interest, therefore, are a consequence of the spatial distribution of the charges rather than their movement *per se*.

In the following then, we begin with a brief description of electrical charges and from this develop concepts such as electric field, potential and capacitance. Some basic properties of insulators are introduced but a detailed description of the mechanisms by which static charges are generated in industry is left until the next chapter.

## 2.1  ELECTROSTATIC CHARGES

All matter consists of microscopic entities known as atoms. These are further subdivided into a central nucleus around which electrons orbit. The nucleus, which contains most of the mass of the atom, is made up from a hierarchy of elementary particles. We need consider only the proton which carries a charge of $+1.6 \times 10^{-19}$ C. Normally, atoms are electrically neutral because, for each positively charged proton in the nucleus, an electron carrying a negative charge of $-1.6 \times 10^{-19}$C orbits the nucleus so that the overall charge on the atom is zero. This balance can be upset in many atomic species since electrons can transfer readily from one type of atom to another. For example, in ionic solutions a diatomic molecule, AB, may dissociate as follows

$$AB \rightarrow A^+ + B^-$$

generating a positive ion, $A^+$, and a negative ion, $B^-$. High energy radiation such as UV light, x-rays or radioactivity can produce the same result and may cause even relatively stable and more complex molecules to dissociate into charged fragments, e.g.

$$ABC \rightarrow AB^+ + C^-.$$

Intense electric fields can cause similar effects. In addition to being charged, any molecular fragments produced by such mechanisms are also likely to be very reactive chemically.

Finally, it should be noted that electrons can simply transfer from one surface to another as a result of a contact potential difference (or electrochemical potential difference) between them. Consequently, the charge balance in both surfaces is disturbed; one surface charging positively, the other negatively.

As we will see later, it is the separation of positive and negative charges at interfaces between different materials that usually leads to unwanted static electricity in industry.

## 2.2  FORCES BETWEEN CHARGES

Electrical charges exert forces on each other. A systematic investigation of these forces reveals that there must be two types of charge, positive and negative, in line with our model of the atom. Similar charges, say two positives, repel each other while dissimilar charges, one positive and one negative, attract each other. These effects are reminiscent of the forces exerted by the poles of a bar magnet and indeed there is a strong analogy between magnetic and electrostatic effects.

## 2.2.1 Coulomb's Law

In 1785 Coulomb postulated that the magnitude of the force between two electrical charges depends on the magnitudes of the charges themselves and on their spatial separation. By analogy with magnetism, and after extensive experimentation, he suggested that if a point charge $+ q'$ is located at a distance r from a positive point charge, $+ q$, then a force, F, acts along the line joining the centres of the charge as shown in Figure 2.1.

+q                                    +q'                   F

**Figure 2.1**   *Direction of electrostatic force acting on a point charge +q' in the field of a point charge +q.*

The magnitude of the force is given by the inverse square law, i.e.

$$F = \frac{qq'}{4\pi\varepsilon\varepsilon_0 r^2} .$$                                    (2.1)

In the SI system, the unit of charge is the Coulomb and $\varepsilon_0$, the absolute permittivity of free space, has the value $8.85 \times 10^{-12}$ F/m. The relative permittivity (or dielectric constant), $\varepsilon$, accounts for the presence of the medium in which the charges are located. From equation (2.1), therefore, the force exerted by a charge of $+1$ μC on a charge of $+0.01$ μC located 10 mm away in air ($\varepsilon \sim 1$) is one of repulsion and has a magnitude of

$$F = \frac{10^{-6} \times 10^{-8}}{4\pi \times 8.85 \times 10^{-12} \times 10^{-4}}$$

$$F = 0.9 \text{ Newton.}$$

Examples of these forces are frequently seen in industry. The cling of plastic films, the billowing of yarn, the dust pick-up seen on charged surfaces are all the result of coulomb forces between electrostatic charges.

## 2.3 ELECTRIC FIELD

The existence of Coulomb forces between charges leads to the first important concept in electrostatics, namely, the electric field. This is simply a region of space in which electrical (or coulomb) forces act. Therefore, every charged object is surrounded by an electric field.

The field is visualised by drawing lines of force to indicate the direction in which the force acts (Figure 2.2). These lines show the path that would be followed by a small, weightless, positive point charge placed in the field. Consequently, lines of force must always commence on a positive charge and terminate on a negative charge. In the case of the single point charge in Figure 2.2(a) the lines of force are all directed radially outwards, implying that the negative charge on which they terminate is distributed uniformly over the surface of a sphere of infinite radius. In Figure 2.2(b) the simple radial symmetry is lost because other charges have now been introduced into the vicinity of the original point charge.

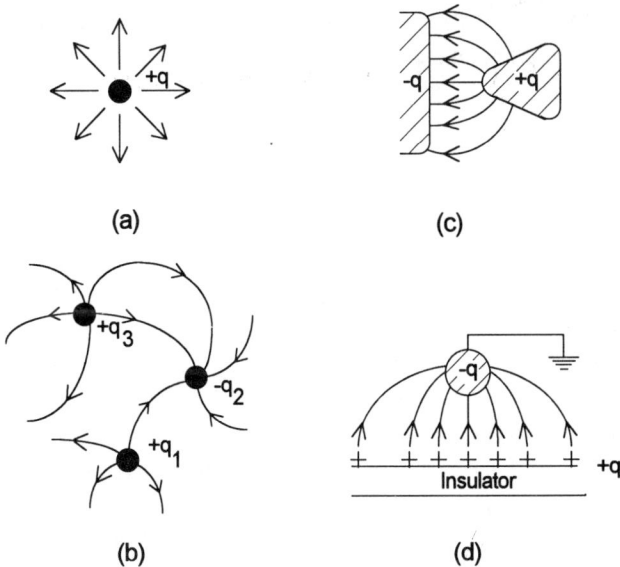

**Figure 2.2**    *Examples of electric fields in the vicinity of (a) a positive point charge (b) near point charges of various magnitudes and polarity (c) between two oppositely charged irregularly shaped conductors and (d) between a positively charged insulating sheet and an earthed conductor.*

In the early stages of electrostatic theory it is usual to consider only point charges but the arguments apply equally well to charges spread over the surfaces of conductors and insulators. In Figure 2.2(c) the lines of force linking two charged but irregularly shaped conductors are shown. Electric fields can be set up also between a charged insulator and an earthed conductor as shown in Figure 2.2(d).

The strength of an electric field is defined in terms of the force that an imaginary test charge would experience if it were placed in the field. Thus, the magnitude, E, of an electric field at a particular location in an electrostatic system is the force acting on a unit positive charge placed at that point. Thus, the force experienced by a point charge, $q$, in a field E is

$$F = qE.    \qquad (2.2)$$

### 2.3.1  Electric Field of a Point Charge

From the above definition, the strength of the electric field surrounding the point charge, $+q$, in Figure 2.2(a) is readily  calculated. Imagine that a charge of $+ 1$ C is positioned at some distance r from the charge $+q$. From Coulomb's law (equation (2.1)), the force acting on the unit charge and, therefore, the strength of the electric field at that point are both given by

$$E(r) = \frac{q}{4\pi\varepsilon\varepsilon_0 r^2}.    \qquad (2.3)$$

The field will have the same strength at all points on the surface of an imaginary sphere of radius, r, drawn around, and concentric with, the charge. Furthermore, the field will be directed radially away from the charge as indicated by the lines of force in Figure 2.2(a).

### 2.3.2  Flux Density (Displacement)

Where lines of force are used to represent the magnitude of the charge from which they emanate, they are termed flux lines and, by definition, $q$ lines of force are assumed to emanate from a charge $+q$ Coulombs.Symmetry arguments dictate that all flux lines must emanate uniformly from a point charge. So for our charge $+q$, the density D(r) of flux lines passing through the surface of an imaginary, concentric sphere of radius r surrounding the charge is given by

$$D(r) = \frac{q}{4\pi r^2} = \varepsilon\varepsilon_0 E(r).    \qquad (2.4)$$

The flux density is often referred to as the displacement and from equation (2.4) it can be seen that it is directly proportional to the electric field at that point. Where the flux density is high, the electric field is high and where it is low, the electric field is low. In Figure 2.2(a) we see that the fields are strongest close to the point charge and decrease further away in accordance with equation (2.3).

In Figure 2.2(c) and (d) the flux densities and therefore the electric fields are seen to be strongest where the charged surfaces are most curved. In other words, sharply curved surfaces significantly enhance the electric fields in an electrostatic system.

This important principle is exploited in most types of static neutralisers, where sharp points are used to enhance electric fields locally in order to induce a partial breakdown of the air in the vicinity of the points. This partial breakdown of the air is termed a corona because of the "halo" it produces around the point. The ions formed in the discharge are used to neutralise static charges generated on products and parts of the production system during manufacture.

The principle is applied in reverse when designing high-voltage terminals and conductors where electrical stress relief and the suppression of corona discharges are achieved either by forming such structures from components with large radii or enclosing them in conducting spheres of large radius.

### 2.3.3  Electric Field of an Infinite Sheet of Charge

The concepts developed above for point charges may easily be applied to more extensive charge distributions. For example, we may assume that the charged sheet in Figure 2.3 is composed of an infinite array of point charges. From the principle of superposition, the electric field of the charged sheet is readily calculated, since it is equal to the sum of the electric fields from every point charge in the array.

Assume that the charge density on the sheet is $Q$ C/m$^2$. Imagine that an annulus of radius, r, and width, dr, is drawn on the surface of the sheet. Consider now the element of area on this annulus bounded by the two radii making angles of $\phi$ and $\phi+d\phi$ with the horizontal. If the angle $d\phi$ and the width of the annulus, dr, are sufficiently small the area of the element, equal to r.d$\phi$.dr, will also be very small, and the charge, $Q$.r.d$\phi$.dr, residing on this element of area may be thought of as a point charge. The electric field at a point, P, lying at a perpendicular distance, x, from the sheet, is easily calculated from Coulomb's law by assuming that the sheet of charge consists of an infinite number of such charges. Thus the contribution $E_\phi(x)$ to the electric field at P from the element with coordinates $(r,\phi)$ on the sheet is given by

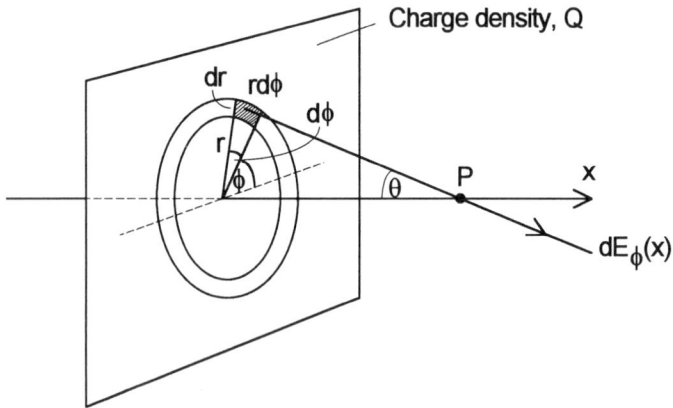

**Figure 2.3** *Contribution to the electric field from an element of area on a charged sheet.*

$$dE_\phi(x) = \frac{Q.r.dr.d\phi}{4\pi\varepsilon\varepsilon_0(r^2 + x^2)} \qquad (2.5)$$

and directed at an angle $\theta$ to the x-axis as shown in Figure 2.3. The annulus is made up of $2\pi$ such elements so that the field, $dE_p(x)$, at P from the annulus is

$$dE_P(x) = \frac{Q.2\pi r.dr}{4\pi\varepsilon\varepsilon_0(r^2 + x^2)} \cos\theta$$

i.e. $$dE_P(x) = \frac{Qx}{2\varepsilon\varepsilon_0} \times \frac{rdr}{(r^2 + x^2)^{\frac{3}{2}}} \qquad (2.6)$$

Note that it is only necessary to consider the component of field normal to the sheet because the sum of the parallel contributions vanishes. The total field at P is now obtained by summing the contributions from each annulus that makes up the sheet. Thus,

$$E_P(x) = \frac{Qx}{2\varepsilon\varepsilon_0} \int_0^\infty \frac{r.dr}{(r^2 + x^2)^{\frac{3}{2}}}$$

which upon integrating yields

$$E_P(x) = \frac{Q}{2\varepsilon\varepsilon_0}. \tag{2.7}$$

This most interesting and useful result leads to the following conclusions:

(i) The electric field surrounding an infinite, uniformly charged sheet has the same magnitude everywhere and is directed normally to the sheet.

(ii) The flux from the sheet is divided equally between the two sides so that the electric fields from the sheet are in opposite directions on the two sides of the sheet.

The breakdown field of air, $E_{BD}$, is normally taken to be 2.7 MV/m, so the maximum charge density, $Q_{MAX}$, that can be tolerated on a large, isolated plastic sheet is given by

$$Q_{MAX} = 2\varepsilon\varepsilon_0 E_{BD} = 2 \times 1 \times 8.85 \times 10^{-12} \times 2.7 \times 10^6$$

$$= 4.78 \times 10^{-5} \text{ C/m}^2.$$

If the charging process is so severe that this value of charge density is exceeded then the air adjacent to the sheet breaks down and conducts the excess charge away. Whether this occurs as a harmless low level discharge or a large luminous spark depends on the conditions pertaining at the time.

### 2.3.4 Electric Field of a Volume Charge

When dealing with liquids and powder clouds, electrical charges will be distributed throughout a volume of space usually bounded by the walls of a containing vessel. In a liquid, the charges normally will be carried by ions but in a dust cloud they will be attached to individual powder particles. In both cases though, the system is essentially a collection of point charges and, in principle, the electric field anywhere in the system can be calculated by summing vectorially the electric fields from each point charge. Since the system may contain a large number of such particles (there may be millions of dust particles in a powder cloud), such an approach clearly is not very feasible.

Fortunately, a way to proceed is available to us through the application of Gauss' theorem, which allows the macroscopic effects of a system of microscopic charges to be determined rather simply.

*2.3.4.1  Gauss' Theorem*

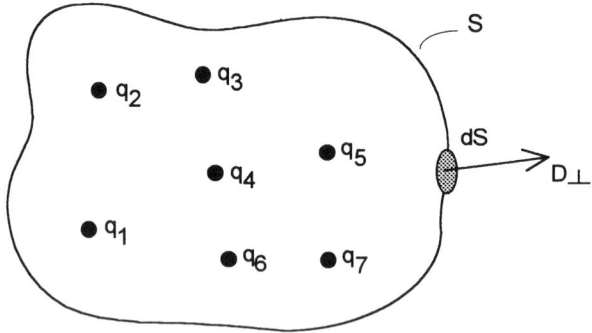

**Figure 2.4**    *General application of Gauss' theorem.*

In essence Gauss' theorem states that if we add together the normal components of the flux emanating from all the small increments of area making up a closed surface, then the sum will be equal to the total charge enclosed by the surface. This is true whatever the shape of the surface and is independent also of the number, magnitude and location of the charges within the enclosed volume. Mathematically, the theorem is written as

$$\oint_S D_\perp dS = Q_{ENCL} \qquad (2.8)$$

where $D_\perp$ is the normal component of the flux density (or displacement) passing through the element of area dS, $Q_{ENCL}$ is the total charge contained within the enclosed volume and the integral is taken over the whole of the enclosing surface, S (see Figure 2.4). The relationship between flux density and electric field has already been given in equation (2.4). Therefore, if the dielectric material in which the charges reside is homogeneous, i.e. the relative permittivity is constant throughout the volume, then equation (2.8) may be written as

$$\varepsilon \varepsilon_0 \oint_S E_\perp dS = Q_{ENCL} \qquad (2.9)$$

where $E_\perp$ is the normal component of the electric field acting at the surface element dS. This equation provides the essential link that we require between the microscopic charged particles represented by $Q_{ENCL}$ and a macroscopic property that can be measured, i.e. the electric field, $E_\perp$. Although the theorem can be applied to closed surfaces of any shape, its power lies in providing a means of solving electrostatic problems associated with highly symmetrical charge distributions. Two practical examples involving a spherical container and a pipe are given below. The calculation of electric fields for more complex geometries is left until section 2.5.

### 2.3.4.2 Electric Fields in Spherical Containers

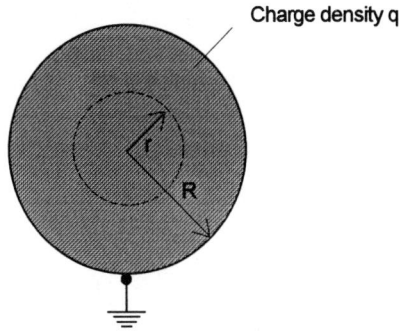

**Figure 2.5**     *Application of Gauss' theorem to a sphere of charge.*

Figure 2.5 shows an earthed, spherical container of radius R in which a uniformly charged dust cloud has been generated so that the volume charge density is q. (The calculation which follows would be equally valid if the container was filled with a uniformly charged liquid). Consider now the imaginary sphere of radius r concentric with the main sphere. The total charge enclosed within the imaginary sphere is $4\pi r^3 q/3$. The spherical symmetry of the charge distribution means that all the flux lines are directed radially outwards from the centre to the walls of the container and must be uniformly distributed over the surface of the sphere. Therefore, the normal component of flux density, $D_\perp(r)$, at the surface of the imaginary sphere must be constant over the whole surface. So, applying Gauss' theorem, equation (2.8), to the imaginary sphere

$$\oint_S D_\perp(r)dS = \frac{4}{3}\pi r^3 q. \qquad (2.10)$$

Since $D_\perp(r)$ is a constant for a given value of r and the integral of dS over the closed surface is simply the surface area of the imaginary sphere, equation (2.10) may be written as

$$D_\perp(r).4\pi r^2 = \frac{4}{3}\pi r^3 q$$

or

$$D_\perp(r) = \frac{rq}{3}.$$

The electric field at r can be obtained from equation (2.4) and is given by

$$E(r) = \frac{rq}{3\varepsilon\varepsilon_0}. \tag{2.11}$$

The analysis above shows that the electric field at the very centre of the sphere (r=0) is zero. This is exactly what is expected because the coulomb forces acting on a unit test charge at the centre will be equal in all directions so that the resultant force will be zero.

The maximum field occurs at the wall of the container and is given by

$$E_{MAX} = E(R) = \frac{Rq}{3\varepsilon\varepsilon_0} \tag{2.12}$$

which is equal to the field that would be observed at the wall if all the enclosed charge were located at the centre of the container. If the charged cloud has been generated in a container of radius 1 m, and the charge density is 1 $\mu C/m^3$, then assuming most of the volume is filled with air ($\varepsilon\sim1$) the strength of the electric field just inside the wall of the container is 37.7 kV/m. If the radius of the container is 10 m, the same charge density would give rise to an electric field of 377 kV/m at the container wall.

This application of Gauss' theorem shows clearly that,by increasing the size of storage tanks without taking steps to reduce the charge density on the dust cloud, the risk of producing an electrical discharge is increased. This conclusion applies to most containers, whatever their shape. The risk can be substantially reduced by dividing the internal volume into a number of "compartments" separated by earthed rails or panels.

### 2.3.4.3 Electric Field in a Pipe

Powder and liquid products transported through pipes will invariably acquire an electrostatic charge. Again, because of the symmetry involved, Gauss' theorem can be used to calculate the magnitude of the electric fields in the pipe.

Figure 2.6 shows a section of a long pipe carrying a uniformly charged product. The pipe radius is R and the volume charge density is q. The imaginary Gaussian surface this time is a cylinder of radius r and length L. The total charge contained inside this surface is $\pi r^2 Lq$. A consequence of the cylindrical symmetry is that all flux lines radiate out uniformly from the axis of the cylinder to the wall, so that $D_\perp(r)$ is constant over the curved surface of the imaginary cylinder. Since all flux lines are normal to the axis of the cylinder, none will cross the end faces of the cylinder. Applying Gauss' theorem as in the previous section, then, yields

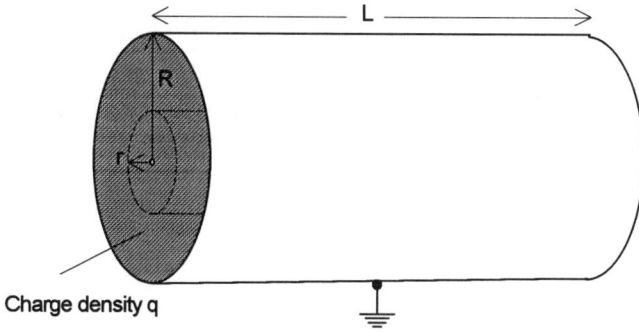

**Figure 2.6**    *Applying Gauss' theorem to charge distributed within a pipe.*

$$D_\perp(r).2\pi rL = \pi r^2 Lq$$

from which

$$D_\perp(r) = \frac{rq}{2}$$

and

$$E(r) = \frac{rq}{2\varepsilon\varepsilon_0}.$$   (2.13)

As in the case of the sphere, the electric field is zero at the centre and is highest just inside the pipe wall. The maximum electric field is given by

$$E_{MAX} = E(R) = \frac{Rq}{2\varepsilon\varepsilon_0}.$$   (2.14)

So, in a long cylindrical pipe of radius 0.3 m containing a hydrocarbon liquid ($\varepsilon \sim 2$) with an associated charge density of 10 $\mu$C/m$^3$, the maximum electric field that can be developed is 84.7 kV/m. Again we see that for the same charge density the maximum field will increase in direct proportion to the pipe radius.

## 2.4 POTENTIAL

The presence of forces between electrical charges means that, if one charge is moved around in the neighbourhood of another, work has to be done by, or on, the moving charge. It follows, therefore, that a stationary electrostatic charge in an electric field must possess potential energy, and if it moves from one location to another in an electric field its potential energy must change. This change in the potential energy is used to define the potential in an electrostatic system.

### 2.4.1 Potential of a Point Charge

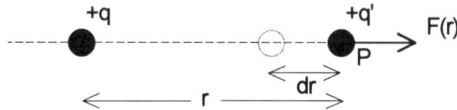

**Figure 2.7**    *Point charge q' moving in the field of a point charge q.*

As an example consider the point charge $+q'$ located at P in Figure 2.7. If it moves a small distance dr towards the point charge $+q$ then work must be done against the force of repulsion it experiences from $+q$. The work done, dW(r), in moving the distance dr is given by

$$dW(r) = -F(r)dr = -\left(\frac{qq'}{4\pi\varepsilon\varepsilon_0 r^2}\right)dr \qquad (2.15)$$

the negative sign appearing because $q'$ is moving to smaller values of r. The total work done, W(r), in bringing the charge $q'$ from infinity to P is then the sum of all such increments. In the limit of dr tending to zero, the summation can be replaced by an integration giving

$$W(r) = -\left(\frac{qq'}{4\pi\varepsilon\varepsilon_0}\right)\int_{\infty}^{r} r^{-2} dr$$

or

$$W(r) = \frac{qq'}{4\pi\varepsilon\varepsilon_0 r}. \qquad (2.16)$$

When $q'$ is a unit test charge then dW(r) is equal to the potential difference between two points distance dr apart and W(r) is defined as the potential at P. So, the potential, V(r), at a point (often referred to as the voltage at that point) is the work done in bringing a unit positive charge from infinity to the point; i.e. at a distance r from a point charge, +q, the potential is given by

$$V(r) = \frac{q}{4\pi\varepsilon\varepsilon_0 r}.$$
(2.17)

From equations (2.3) and (2.17) we obtain the important result that

$$E(r) = -\frac{dV(r)}{dr}.$$
(2.18)

Apart from the minus sign, the electric field at a point is seen to be equal to the potential gradient at that point. Although this result has been derived for the case of point charges, it is generally applicable and written as

$$E(x,y,z) = -\nabla V(x,y,z)$$
(2.19)

where $\nabla$ represents the partial differentials $\left(\frac{\partial}{\partial x} + \frac{\partial}{\partial y} + \frac{\partial}{\partial z}\right)$ in the Cartesian coordinate system. By arbitrarily assuming that the potential energy of a charge at infinity is zero, then equations (2.16) and (2.17) give the potential energy, U(r), of a point charge $q'$ located at a point where the potential is V(r) as

$$U(r) = q'V(r).$$
(2.20)

Therefore, a point charge of 1 μC raised to a potential of 10 kV has 10 mJ of potential energy. If the electric field at that same location is 1 MV/m a force of 1 Newton acts on the charge.

### 2.4.2 Potential of an Infinite Sheet of Charge

In section 2.3.3 it was shown that the electric field of an infinite, uniform sheet of charge was directed normally to the sheet and at all locations had a constant value $Q/2\varepsilon_0$ in air, where $Q$ is the charge density on the sheet. If the sheet is parallel to the yz plane, then from equation (2.18) the potential is a linear function of x only since

$$\int dV(x) = -\int E(x)dx$$

or

$$V(x) = -\frac{Qx}{2\varepsilon_0} + V_R$$

where $V_R$ is a constant of integration. The actual value of $V_R$ is immaterial because in electrostatic problems we will always be concerned with the *difference* in potential between two locations. Thus the potential difference, $V_{12}$, between two points located at distances $x_1$ and $x_2$ away from a uniformly charged sheet is given by

$$V_{12} = \frac{Q}{2\varepsilon_0}(x_2 - x_1) \ .$$

If $Q = 10 \ \mu C/m^2$, $x_1 = 2$ cm and $x_2 = 5$ cm then the potential difference between the two points is 16.9 kV.

### 2.4.3  Equipotentials

It should be obvious that the expressions derived above for potential can be used to draw equipotentials (lines or surfaces of constant potential) around the charges in a system. Figure 2.8(a) and (b) show respectively a set of equipotentials (dashed lines) for a point charge and an infinite sheet of charge. As can be seen, and in accordance with equation (2.18), the equipotentials are everywhere orthogonal (at right angles) to the electric field lines.

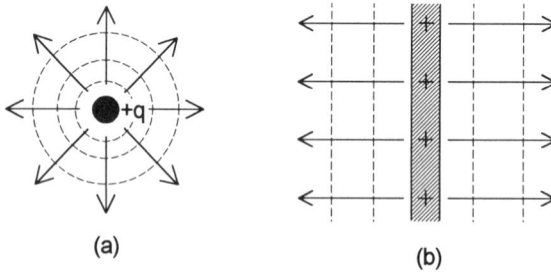

(a)                                         (b)

**Figure 2.8**     *Electric field (———) and equipotential (- - -) lines near (a) a point charge and (b) a charged sheet.*

### 2.5  FIELDS AND POTENTIALS ASSOCIATED WITH A VOLUME CHARGE

Apart from the few highly symmetrical charge distributions discussed above, Gauss' theorem in its original form cannot be applied directly to real systems.

However, the theorem can be developed into two particularly useful equations which, when solved with appropriate boundary conditions, yield solutions to all electrostatic field problems.

Consider a volume of space containing a charge density q(x,y,z), whose magnitude can change from position to position. Now draw a Gaussian surface to enclose a small element of this volume. As shown in Figure 2.9 the Gaussian surface is arranged to be a cuboid whose sides dx, dy and dz lie parallel with the x, y and z axes respectively and are sufficiently small that the charge density is constant within the volume element. The displacement D(x,y,z) entering this volume can be divided into three components, $D_x$, $D_y$ and $D_z$, parallel to the x, y and z axes respectively.

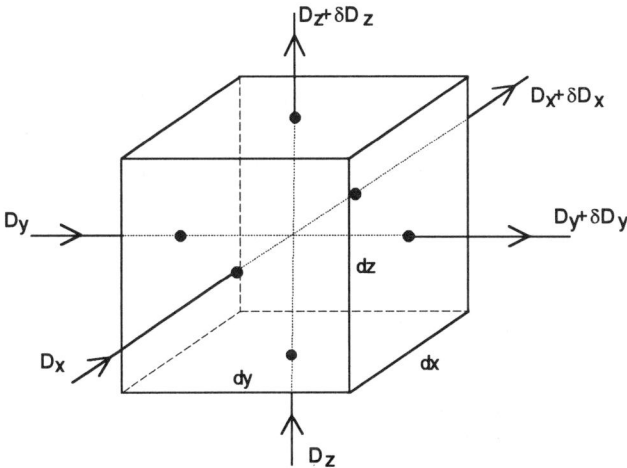

**Figure 2.9**    *Application of Gauss' theorem to this element of volume yields Poisson's equation.*

The components of the displacement leaving the volume element are $D_x+\delta D_x$, $D_y+\delta D_y$ and $D_z+\delta D_z$. From Gauss' theorem, the differences in these components of the displacement must arise from the presence of charge in the volume. So, from equation (2.8) we may write

$$\delta D_x\, dydz + \delta D_y\, dxdz + \delta D_z\, dxdy = q\, dxdydz \qquad (2.21)$$

Furthermore, if the sides of the cuboid are sufficiently small, then

$$\delta D_x = \frac{\partial D(x,y,z)}{\partial x} dx, \quad \delta D_y = \frac{\partial D(x,y,z)}{\partial y} dy, \quad \delta D_z = \frac{\partial D(x,y,z)}{\partial z} dz.$$

Substituting these relations into equation (2.21) and dividing through by the volume of the cuboid, dx.dy.dz, yields

$$\left(\frac{\partial}{\partial x} + \frac{\partial}{\partial y} + \frac{\partial}{\partial z}\right) D(x,y,z) = q(x,y,z)$$

and from equation (2.4) we obtain

$$\left(\frac{\partial}{\partial x} + \frac{\partial}{\partial y} + \frac{\partial}{\partial z}\right) E(x,y,z) = \frac{q(x,y,z)}{\varepsilon\varepsilon_0} \qquad (2.22)$$

We have shown that if the charge density is known throughout a volume of space then, in principle, the electric fields throughout that volume can be evaluated by solving equation (2.22). We also know that the magnitude of the electric field is equal to the potential gradient. Thus, by combining equations (2.19) and (2.22) the potential V(x,y,z) in the system can be written as

$$\left(\frac{\partial^2}{\partial x^2} + \frac{\partial^2}{\partial y^2} + \frac{\partial^2}{\partial z^2}\right) V(x,y,z) = -\frac{q(x,y,z)}{\varepsilon\varepsilon_0}. \qquad (2.23)$$

This equation is known as the Poisson equation and for convenience is often written as

$$\nabla^2 V = -\frac{q}{\varepsilon\varepsilon_0} \qquad (2.23)$$

where it is understood that both V and q may be functions of position. If no charge is enclosed within the volume under consideration then equation (2.22) reduces to

$$\nabla^2 V = 0 \qquad (2.24)$$

which is the Laplace equation. For many practical systems the geometry is such that analytical solutions of the Poisson or Laplace equations are not possible. In these cases the only recourse is to obtain numerical computer solutions using finite-difference or finite-element techniques (see for example McAllister et al, 1985).

Often though, approximating the system to a sphere or cylinder for which analytic solutions have been presented already in section 2.3.4 can provide a sufficient degree of accuracy to solve particular problems. In many cases also, a one-dimensional model of the system can be solved analytically and can give valuable insight into why an electrostatic problem may have arisen.

Chemical engineers will recognise the Laplace and Poisson equations derived above as the governing equations for fluid flow. It may be helpful for many readers to realise the close analogy that exists between electrostatics and heat flow for which equation (2.23) is written as

$$\nabla^2 T = -\frac{\dot{q}}{\kappa}$$

where T is the temperature, $\dot{q}$ is the rate of heat generation per unit volume and $\kappa$ is the thermal conductivity of the medium. In the Chemical Industry numerical methods are used widely to solve fluid flow problems and a number of software packages are available commercially for this purpose. From the above discussion it should be obvious, therefore, that these packages can be used for solving electrostatic problems also. (It should be noted though that,if the heat flow analogy is used for simulating an electrostatic problem, the magnitudes of the numerical solutions could be very different from those encountered in normal heat flow problems. For example, while potentials of $10^5$ V are readily obtained in an electrostatic system, chemical engineers would not expect to have to deal with temperatures of $10^5$ K !)

## 2.6 BOUNDARIES OF DIELECTRICS

In dealing with electrostatic problems it soon becomes apparent that the bound-aries of dielectrics have rather special properties which have profound effects on the electric fields and potentials in the system. We shall be concerned with charges located near the interfaces between

(i) a dielectric and a conductor, and

(ii) between two dielectrics.

However, before discussing specific cases, we need to introduce some general rules that apply at all dielectric boundaries.

## 2.6.1 General Rules at Dielectric Interfaces

Figure 2.10 shows some of the common situations that arise at dielectric interfaces.

*(a) Electric Fields Normal to the Interface*

In the first case, Figure 2.10(a), an electric field is directed normally to the interface between two semi-infinite slabs of dielectrics. The relative permittivities of the dielectrics are $\varepsilon_1$ and $\varepsilon_2$ respectively. The flux density in material 1 is $D_{N1}$ while that in material 2 is $D_{N2}$, both being normal to the interface. Now apply Gauss' theorem to an imaginary pillbox enclosing a small section of the interface. Since there are no charges enclosed within the pillbox, the flux approaching the interface must always equal the flux leaving,i.e. $D_{N1} = D_{N2}$. Therefore, the normal component of the flux density (displacement) is continuous across a boundary.

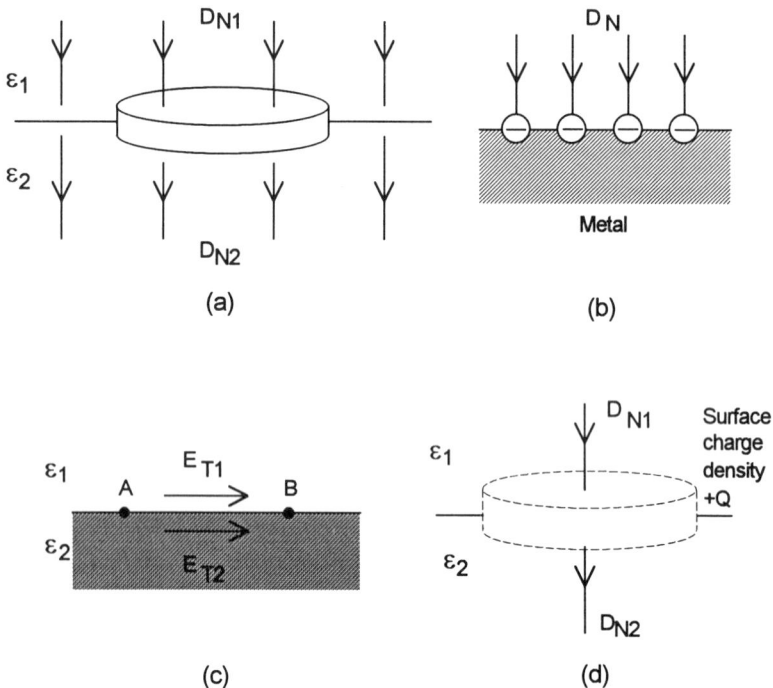

(a)

(b)

(c)

(d)

**Figure 2.10**    *Boundary conditions in electrostatic problems (see text for details).*

Consequently, the ratio of the normal components of electric field across the interface is easily shown to be in the inverse ratio of the relative permittivities of the adjoining materials, i.e.

$$\frac{E_{N1}}{E_{N2}} = \frac{\varepsilon_2}{\varepsilon_1}. \tag{2.25}$$

A good conductor may be assumed to be a material that is infinitely polarisable, i.e. a material whose relative permittivity is infinitely large. If we assume that material 2 is a conductor, then inserting $\varepsilon_2 = \infty$ in equation (2.25) shows that, if $E_{N1}$ is finite, then $E_{N2}$ must be zero. In other words, a good conductor cannot support an electric field and all flux lines must terminate normally on charges located at the surface of the conductor, as shown in Figure 2.10(b). These surface charges on the conductor are said to have been *induced* by the electric field present in the adjacent insulating medium.

### (b) Electric Fields Parallel to an Interface

In Figure 2.10(c) the electric field is directed parallel to the interface between two dielectrics. The magnitudes of the fields in the two materials are $E_{T1}$ and $E_{T2}$ respectively. If we now take a small test charge and move it between two points A and B located on the interface, then the change in its potential energy and therefore the potential difference between A and B should not depend on whether the path of the test charge takes it through material 1 or material 2. The potential at B relative to A is then given by

$$V_B - V_A = -\int_A^B E_{T1}\,dx = -\int_A^B E_{T2}\,dx$$

from which it is deduced that

$$E_{T1} = E_{T2}$$

Since the tangential components of electric field on either side of the boundary are identical, potential must be a continuous function across an interface.

For the case where material 2 is a conductor, we know that $E_{T2}$ must be zero. Therefore, the field $E_{T2}$ in the dielectric close to the interface must also be zero. It follows then that flux lines must be incident normally on a conducting surface.

## (c) Interfacial Charges

When charges are present at the interface between two dielectrics, as in Figure 2.10(d), an extra boundary condition is required which is obtained by applying Gauss' theorem to a pillbox surrounding a small section of the charged interface. Thus, if $Q$ is the charge per unit area of the interface

$$D_{N_2}A - D_{N_1}A = Q.A$$

or
$$D_{N_2} - D_{N_1} = Q. \tag{2.26}$$

On crossing a charged interface, the change in the normal component of flux density is equal to the charge density at the interface.

We can further write that

$$\varepsilon_2 E_{N_2} - \varepsilon_1 E_{N_1} = Q/\varepsilon_0. \tag{2.27}$$

### 2.6.2 Image Charges

As seen above, electric fields have the ability to induce charges in nearby conducting surfaces. Electric fields from positive charges induce negative charges while fields from negative charges induce positive charges. Since the induced charge is always opposite in sign to the original charge then it is readily seen that charged materials such as polymer film, adhesive tape or yarn will always be attracted to conducting structures (including personnel!). Charges are induced in dielectrics as well as in conductors, though in the former the magnitude of the induced charge is smaller and may take some time to reach its equilibrium value. The electric fields and potentials arising from the presence of these induced charges are generally described by invoking the concept of an image charge.

## (a) Image Charges in Conducting Surfaces

The sequence of events leading to the formation of an induced charge in a conducting material is described in Figure 2.11. An isolated, conducting plate is placed at a distance, d, from a point charge $+q$. Initially, the plate will cut across a number of the equipotential surfaces which surround the point charge so that a potential difference and therefore an electric field exists between different positions in the plate. Almost instantaneously the free charges in the conductor, normally electrons, redistribute in such a way as to develop an opposing field. This process is known as polarisation and within a few picoseconds it can proceed to the point where the electric fields in the conductor become zero.

For a plate near a positive charge this will be achieved when an excess of electrons is drawn to the front surface leaving a dearth of electrons (i.e. a positive charge) at the back surface as shown in Figure 2.11(a). When this happens, the surface of the conductor becomes an equipotential, so that all incident flux lines are normal to the surface.

The plate in Figure 2.11(a) will have positive potential because of its proximity to the positive point charge. The plate potential is reduced to zero by earthing, an action which draws negative charges from earth to the plate to neutralise the effects of the positive charges on the back surface. In fact, after earthing, all the positive charges on the back surface of the plate are effectively removed leaving only the negative charges on the front surface (Figure 2.11(b)). Thus a positive charge induces a negative charge at the surface of a conductor and *vice versa*.

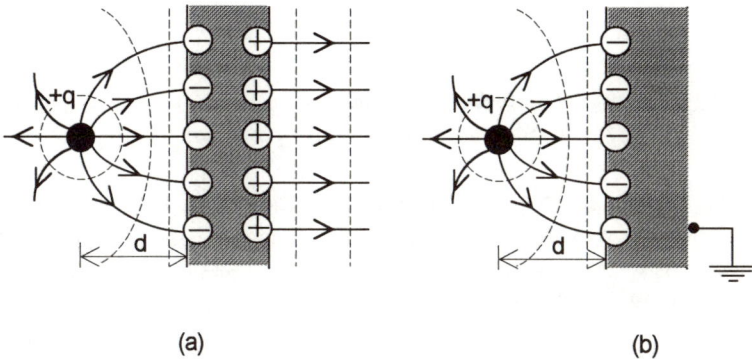

(a)  (b)

**Figure 2.11**    *Inducing charges in metal surfaces. (a) An isolated plate is polarised in the field of a point charge q. (b) When the plate is connected to earth positive charges are conducted away so as to reduce the plate potential to zero.*

Although the negative charges are distributed over the surface of the plate, for mathematical and conceptual ease we imagine that they can all be replaced by a single charge $-q$, located a distance d behind the surface on the extrapolation of the normal from the charge $+q$ to the surface, as shown in Figure 2.12. With this configuration, every point on the conductor surface is equidistant from the two point charges so that the sum of the potentials arising from the two charges is exactly zero everywhere as required. The electric fields and potentials in the space between the charge $+q$ and the plate can be determined by noting that they will be identical to the fields and potentials between two charges $+q$ and $-q$ separated by a distance 2d.

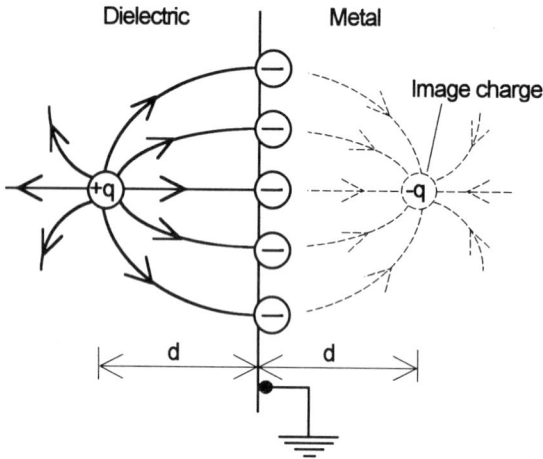

**Figure 2.12**    *The effect of an earthed metal plate near a point charge q can be simulated by replacing the plate with an image charge -q as shown.*

Drawing an analogy with optical reflections, the charge -*q* in Figure 2.12 is known as the image charge. The concept of an image charge is a powerful one and provides us with a method for solving a number of otherwise rather intractable electrostatic problems. For example, if we wish to calculate the force acting on the point charge +*q* placed near a large, flat, earthed plate as in Figure 2.12 we need only consider the force, F, exerted on +*q* by its image charge -*q*. From Coulomb's law (equation (2.1)) this is given by

$$F = \frac{q^2}{4\pi\varepsilon\varepsilon_0(2d)^2} = \frac{q^2}{16\pi\varepsilon\varepsilon_0 d^2}. \qquad (2.28)$$

Unless we invoke the presence of the image charge, then, it would be necessary to sum the forces exerted on +*q* by each negative charge on the surface of the plate. This would require a knowledge of the magnitude and location of every charge on the plate!

*(b) Image Charges in Dielectric Surfaces*
    The formation of an image charge in a dielectric surface follows a similar sequence to that described above for the conductor. However, the lack of free charge carriers in the dielectric means that its response will be limited to the formation of

induced dipoles with an additional contribution from the rotation of permanent dipoles if the dielectric is polar. The tangential field at the interface is unlikely, therefore, to go to zero and the effect of the boundary cannot be simulated with a single image charge.

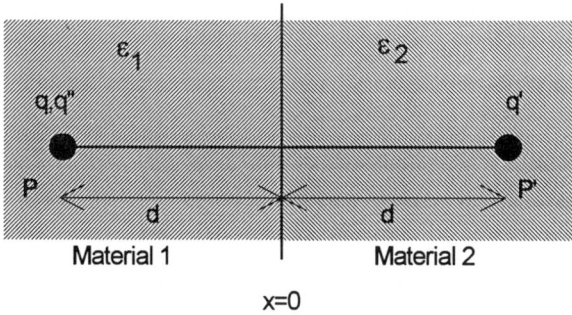

**Figure 2.13**    *Image charges at the interface between two dielectrics.*

Smythe (1968) and Nayfeh and Brussel (1985) show that, for a point charge near a dielectric boundary (Figure 2.13), two image charges are required to describe the field and potentials in the two dielectrics. One image, $q'$, is at the same image position, P', as for a conductor, and together with the original charge, $q$, defines the potentials in material 1, i.e. for $x < 0$ the potential is given by

$$V_1(x,y,z) = \frac{1}{4\pi\varepsilon_1\varepsilon_0}\left(\frac{q}{r}+\frac{q'}{r'}\right) \tag{2.29}$$

where $r = \left[(x+d)^2+y^2+z^2\right]^{\frac{1}{2}}$ and $r' = \left[(x-d)^2+y^2+z^2\right]^{\frac{1}{2}}$. The second image charge, $q''$, is located at P, the position of the original point charge, but this one is used only for defining the potential in material 2. So for $x>0$

$$V_2(x,y,z) = \frac{1}{4\pi\varepsilon_2\varepsilon_0}\frac{q''}{r}. \tag{2.30}$$

At the boundary, $x=0$, both expressions must yield the same potential. Therefore,

$$\varepsilon_2(q+q') = \varepsilon_1 q''. \tag{2.31}$$

Noting also that the normal component of flux density is continuous across the boundary (section 2.6.1) then at $x = 0$

$$\varepsilon_1 \frac{\partial V_1(x,y,z)}{\partial x} = \varepsilon_2 \frac{\partial V_2(x,y,z)}{\partial x}$$

so that

$$q - q' = q''. \tag{2.32}$$

Solving equations (2.31) and (2.32) simultaneously for the image charges yields

$$q' = \frac{\varepsilon_1 - \varepsilon_2}{\varepsilon_1 + \varepsilon_2} q \tag{2.33a}$$

and

$$q'' = \frac{2\varepsilon_2}{\varepsilon_1 + \varepsilon_2} q \tag{2.33b}$$

We see now that if $\varepsilon_1 < \varepsilon_2$ the image charge $q'$ is opposite in sign to the original charge, $q$, which must, therefore, be attracted towards the interface as in the case of a conducting surface. However, the force of attraction is reduced in the ratio $(\varepsilon_1 - \varepsilon_2)/(\varepsilon_1 + \varepsilon_2)$. Interestingly, if $\varepsilon_1 > \varepsilon_2$ the charge $q$ experiences a force repelling it from the interface because the image charge now has the same polarity as $q$.

### 2.6.3 Method of Images

The technique of using image charges for solving electrostatic problems is known as the Method of Images and can be applied to a range of different situations in addition to those described above. Indeed, recognising a suitable image charge or set of charges is an important starting point for many electrostatic problems.

*(a) Line Charge Parallel to an Earthed Plane*

A line charge may be assumed to be constructed from a series of point charges. Therefore, the image of a positive line charge of strength $+\lambda$ per unit length, distance d from an earthed conducting plane, is a negative line charge of strength $-\lambda$ per unit length located at a distance d behind the surface and directly below the real charge as in Figure 2.14.

The magnitude of the electric field acting on the positive line charge is $(\lambda/4\pi\varepsilon\varepsilon_0 d)$ and equal to that arising from a negative line charge placed at a distance 2d away. The force, F, per unit length on the line charge, therefore, is given by

$$F = \frac{\lambda^2}{4\pi\varepsilon\varepsilon_0 d}. \tag{2.34}$$

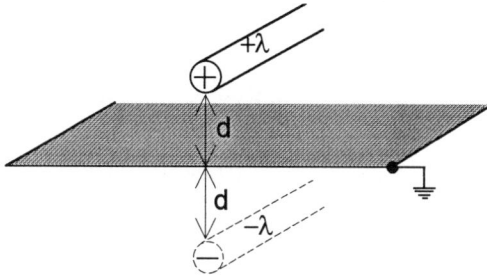

**Figure 2.14**    *Image of a line charge in a conducting plane.*

Equation (2.34) can be used to calculate the force of attraction on any electrostatically charged fibre lying with its axis parallel to an earthed, conducting plane.

*(b) Sheet of Charge Parallel to an Earthed Plane*

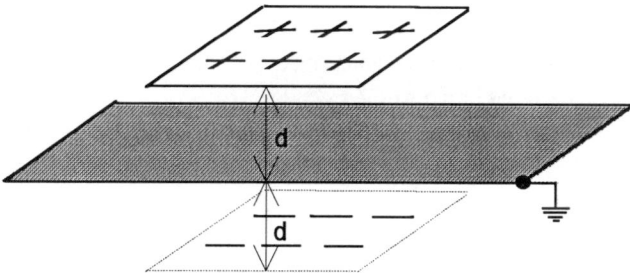

**Figure 2.15**    *Image of a charged sheet in a conducting plane*

A charge sheet can also be assumed to be formed from an array of point charges. Consequently, the image of a positive charge sheet lying parallel to and at a distance d from an earthed, conducting plane is a negative charge sheet located the same distance d below the surface of the plane (Figure 2.15). The electric fields and potentials resulting from a uniformly charged sheet have been dealt with in sections 2.3.3 and 2.4.2. The magnitude of the electric field acting on the real charge is $Q/2\varepsilon\varepsilon_0$ (equation (2.7)) which results in a force, F, per unit area, attracting the sheet towards the surface. The magnitude of F is given by

$$F = \frac{Q^2}{2\varepsilon\varepsilon_0}. \qquad (2.35)$$

The upper surface of an insulating sheet lying on an earthed conducting plane can be charged to a limiting value determined by the breakdown strength of the insulating material, typically in the range 100 - 500 MV/m. Using equation (2.4) we may rewrite equation (2.35) as

$$F = \tfrac{1}{2}\varepsilon\varepsilon_0 E_I^2 \qquad (2.36)$$

where $E_I$ is the field inside the insulator. Assuming that $\varepsilon = 2$, $E_I = 100$ MV/m, the force of attraction per unit area (i.e. presssure) holding the sheet to the conductor is

$$F = \tfrac{1}{2} \times 2 \times 8.85 \times 10^{-12} \times (100 \times 10^6)^2$$

$$= 8.85 \text{ x } 10^4 \text{ N/m}^2$$

i.e. equivalent to almost 9 atmospheres!

*(c) Image of a Charged Sheet Tangential to an Earthed Roller*

The image of a line charge $+\lambda$ in an earthed cylinder is a line charge $-\lambda$ lying on the plane joining the real charge to the cylinder axis. For a real charge distance d from the axis of a cylinder of radius, R, the image charge is located a distance $R^2/d$ from the axis as shown in Figure 2.16(a). The force of attraction per unit length of the line charge is given this time by

$$F = \frac{\lambda^2}{2\pi\varepsilon\varepsilon_0\left(d - \frac{R^2}{d}\right)}. \qquad (2.37)$$

A uniform sheet of charge can be simulated by a series of identical line charges, each with an image in the cylinder. Using the geometrical relations in Figure 2.16(a) it is easily shown that,when the charged sheet is tangential to an earthed cylinder, the image charges are to be found on the surface of a cylinder of radius R/2, lying inside the earthed cylinder as shown in Figure 2.16(b),which, of course, can be used to simulate the transport of insulating film product over rollers (Horvath and Berta, 1975). In principle, both the electric fields and potentials surrounding the roller may be determined from the interaction of the real charge with its image in the roller. The presence of image charges in the earthed roller is the reason, of course, that a charged web tends to snap back to the roller if tension is lost.

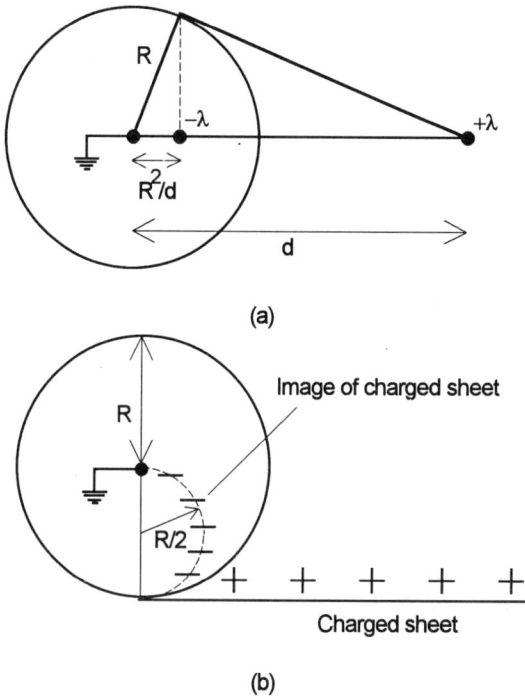

(a)

(b)

**Figure 2.16**    *Construction for locating (a) a line charge and (b) a charged sheet in a conducting roller.*

## 2.7 CAPACITANCE

A particularly important concept in electrostatics is the capacitance of a system. Capacitance quantifies the ability of a system to store electrostatic charge and, as will be shown later, quantifies also the electrical energy that can be stored in the system. All electrostatic systems including charged plastic sheets and dust clouds have an associated capacitance. However, the term is most familiar when applied to a pair of isolated conductors such as the plates of the electronic component known as the capacitor shown in Figure 2.17. The capacitance, C, of such components is defined as the charge that must be transferred from one conductor to the other to produce a unit change in the potential between the conductors.

Therefore, if a charge, q, transferred from one conductor to the other produces a potential difference, V, between them, then the capacitance is given by

$$C = \frac{q}{V}. \tag{2.38}$$

If two conductors are intially at the same potential and a charge of 0.1 μC transferred from one to the other produces a potential difference of 1 kV between the conductors, then the capacitance of the system is $10^{-10}$ F (or 100 pF).

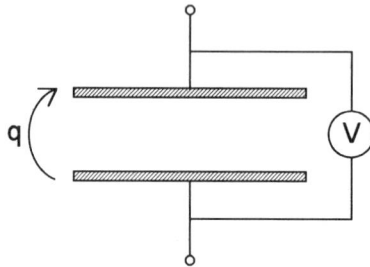

**Figure 2.17**    *A capacitor formed from two isolated conductors.*

### 2.7.1 Parallel-Plate Capacitor

The parallel-plate capacitor is the simplest capacitor system to deal with theoretically, and although not all electrostatic systems can be represented by such a capacitor, nevertheless, the general forms of the relationships derived below are similar in all systems.

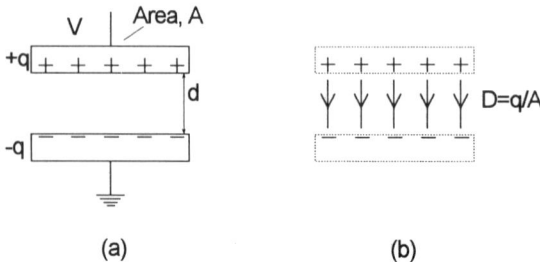

(a)                                    (b)

**Figure 2.18**    *(a) A parallel-plate capacitor and (b) two parallel charge sheets.*

Figure 2.18(a) shows a parallel-plate capacitor consisting of two metal plates of area A separated by a distance d. In transferring a charge $+q$ from the lower plate to the upper, a charge $-q$ is left on the lower plate and a potential difference V appears between the plates. Except near the edges of the plates where fringing effects occur, the charge will distribute itself uniformly over the area of the plate. (This is ensured by the normal conduction process which occurs in metals). If the plate separation, d, is small compared with the lateral dimensions of the plates then the capacitor may be considered as two parallel, uniformly charged sheets as shown in Figure 2.18(b). The charge density on the upper sheet is $+q/A$ and on the lower sheet, $-q/A$, and since no other charges are present in the system, the electric field in the space between the plates has a magnitude $(q/\varepsilon\varepsilon_0 A)$ and is directed towards the lower plate.

If the potential of the lower plate is zero, the potential of the upper plate can be determined from equation (2.18). Thus,

$$V = -E.d = \frac{qd}{\varepsilon\varepsilon_0 A}$$

from which

$$q = \frac{\varepsilon\varepsilon_0 A}{d} V. \tag{2.39}$$

From equations (2.38) and (2.39) it is seen that the capacitance of a parallel-plate capacitor is given by

$$C = \frac{\varepsilon\varepsilon_0 A}{d}. \tag{2.40}$$

We see, therefore, that the capacitance of a parallel-plate capacitor is directly proportional to the area of the plates and inversely proportional to the plate separation. The influence of the medium between the plates is accounted for by its relative permittivity, $\varepsilon$.

When the shape of the capacitor departs from the parallel-plate geometry in Figure 2.18, the capacitance of the system differs from equation (2.40) only in the exact form of the geometry factor A/d. Nevertheless, as will be shown below, large surface areas and small separation between conductors always lead to large values of capacitance.

### 2.7.2 Cylindrical and Spherical Capacitors

Many items of industrial plant have shapes that approximate to cylinders or spheres so that their capacitances may be estimated theoretically to a reasonable degree of accuracy using the formulae derived below. It must be realised, though, that the models analysed represent ideal situations. In practice, the capacitances of such objects may be considerably higher because of the effects of stray capacitance to and from other parts of the supporting structure. Nevertheless, the formulae given are useful for estimating the capacitance of structures, particularly where measurements would be impracticable.

*(a) Coaxial Cylinders*
   The capacitance of a pair of concentric cylinders is easily calculated using Gauss' theorem. Assume that a charge $+\lambda$ per unit length has been transferred from the earthed outer conductor in Figure 2.19 to the inner conductor, thus raising

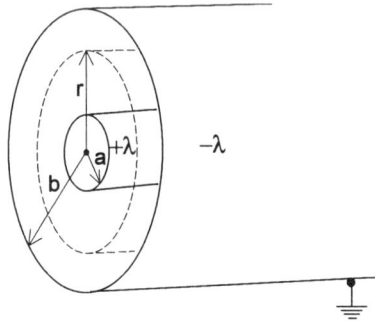

**Figure 2.19**   *A capacitor formed from coaxial cylinders.*

its potential to $V$. A coaxial, cylindrical Gaussian surface of radius r, drawn around the inner conductor, will enclose a charge $+\lambda$ per unit length. Assuming end effects can be neglected and taking note of the symmetry of the capacitor, the electric field will radiate uniformly from the axis. Applying Gauss' theorem (equation (2.9)) it is easily shown that

$$E(r) = \frac{\lambda}{2\pi\varepsilon\varepsilon_0 r}. \qquad (2.41)$$

The potential difference between the conductors is readily obtained since

$$V = -\frac{\lambda}{2\pi\varepsilon\varepsilon_0} \int_b^a \frac{dr}{r} = \frac{\lambda}{2\pi\varepsilon\varepsilon_0} \ln\left(\frac{b}{a}\right). \qquad (2.42)$$

From the definition of capacitance given earlier, we see that the capacitance per unit length of coaxial cylinders is given by

$$C = \frac{2\pi\varepsilon\varepsilon_0}{\ln(b/a)}.$$  (2.43)

*(b) Parallel Cylinders*

Gauss' theorem may also be used for determining the capacitance of a pair of parallel cylinders, but only for the case where the radii of the cylinders are small compared with their separation, as in Figure 2.20.

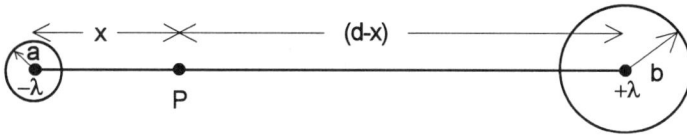

**Figure 2.20**   *Capacitance of parallel cylinders.*

Again we transfer a charge $+\lambda$ per unit length from one conductor to the other. The electric field at P is the result of two forces - attraction to the negative charge and repulsion from the positive charge, and from Gauss' theorem its magnitude is given by

$$E(x) = \frac{-\lambda}{2\pi\varepsilon\varepsilon_0 x} - \frac{+\lambda}{2\pi\varepsilon\varepsilon_0(d-x)}$$

$$= -\frac{\lambda}{2\pi\varepsilon\varepsilon_0}\left(\frac{1}{x} + \frac{1}{d-x}\right).$$

Integrating from a to (d-b) gives the potential difference between the cylinders as

$$V = \frac{\lambda}{2\pi\varepsilon\varepsilon_0}\ln\left(\frac{d^2}{ab}\right)$$

so long as $a\approx b\ll d$. The capacitance of the cylinders is then

$$C = \frac{2\pi\varepsilon\varepsilon_0}{\ln(d^2/ab)}.$$  (2.44)

When the radii of the cylinders are comparable with their separation, then this simple application of Gauss' theorem fails because the electric field of the two cylinders cannot be assumed to arise from line charges located on their axes. However, the method of images (section 2.6.3) shows that the cylinders can still be represented by line charges, but that their positions must satisfy the geometric relationships in Figure 2.21. Using this approach, Smythe (1968) shows that the capacitance per unit length of two parallel cylinders is given by

$$C = 2\pi\varepsilon\varepsilon_0 / \cosh^{-1}\left(\frac{d^2 - a^2 - b^2}{2ab}\right). \qquad (2.45)$$

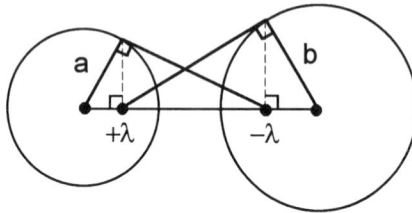

**Figure 2.21**    *Construction for calculating the capacitance between two closely spaced parallel cylinders.*

The capacitance between two identical cylinders is obtained by letting a=b=R. The resulting equation can also be written (Cross, 1987) as

$$C = \pi\varepsilon\varepsilon_0 / \ln\left[\frac{d + (d^2 - 4R^2)^{\frac{1}{2}}}{2R}\right].$$

*(c) Cylinder and a Plane*

The method of images was also used by Nayfeh and Brussel (1985) to determine the capacitance per unit length of a conducting cylinder of radius R whose axis lies parallel to an infinite conducting plane (Figure 2.22). If the axis of the cylinder is at a distance D from the plane, the capacitance per unit length is given by

$$C = \frac{2\pi\varepsilon\varepsilon_0}{\cosh^{-1}(D/R)}. \qquad (2.46)$$

The same result is obtained from equation (2.45) by letting d = D + b and assuming that b = ∞.

The above analysis shows quite clearly that any unearthed metal pipes in a plant will have an associated capacitance. As an example, consider a short metal section used for connecting plastic pipework as in Figure 2.23. If the metal section is 0.5m long, has a diameter of 0.2 m and its axis is located 0.3 m above the ground, equation (2.46) shows that it will have a capacitance of about 16 pF. This calculation will underestimate the true value, which will include the effects of stray capacitances from other earthed metalwork.

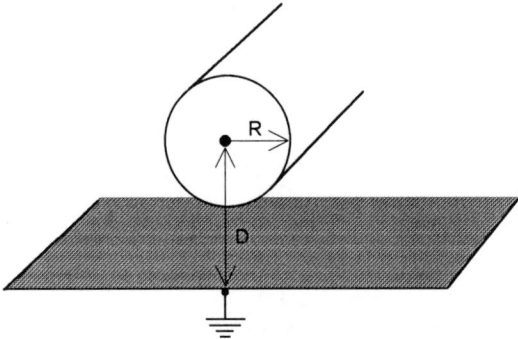

**Figure 2.22**    *Capacitor formed from a cylinder lying above a conducting plane.*

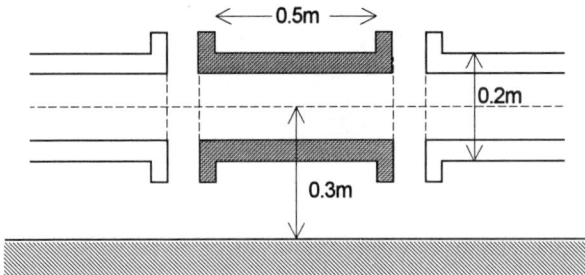

**Figure 2.23**    *A metallic section connecting two plastic pipes has a capacitance to earth.*

*(d) Concentric Spheres*

The capacitance of two conducting, concentric spheres is readily determined from Gauss' theorem. Consider the situation in Figure 2.24 where a charge +q has been transferred from the earthed outer sphere to the inner sphere raising its potential to $V$. From symmetry considerations, the electric field at the surface of an imaginary sphere of radius r enclosing the inner conductor is given by

$$E(r) = \frac{q}{4\pi\varepsilon\varepsilon_0 r^2}.$$

The potential of the inner sphere relative to the outer is given by

$$V = -\frac{q}{4\pi\varepsilon\varepsilon_0}\int_b^a \frac{dr}{r^2} = \frac{q}{4\pi\varepsilon\varepsilon_0}(1/a - 1/b)$$

from which the capacitance is seen to be given by

$$C = \frac{4\pi\varepsilon\varepsilon_0}{(1/a - 1/b)}.$$    (2.47)

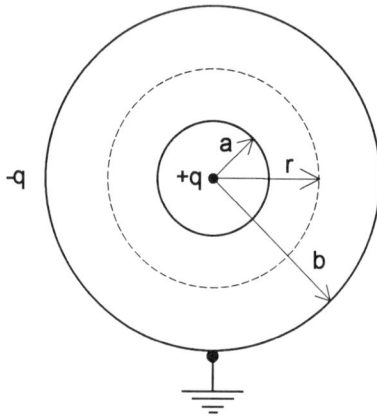

**Figure 2.24**    *Concentric conducting spheres.*

If the radius of the outer sphere is much larger than the inner, then equation (2.47) gives the familiar equation for the capacitance of an isolated sphere, namely

$$C = 4\pi\varepsilon\varepsilon_0 a.$$    (2.48)

Equation (2.48) shows that accurately machined spheres suspended well away from all other structures can be used as standards for calibrating capacitance measuring instruments.

*(e) Sphere and a Plane*
    Smythe (1968) shows that the capacitance between two nonconcentric spheres (Figure 2.25) is given by

$$C = 4\pi\varepsilon\varepsilon_0 ab \sinh a \sum_{n=1}^{\infty} \frac{1}{b \sinh n\alpha + a \sinh (n-1)\alpha} \quad (2.49)$$

where

$$\alpha = \cosh^{-1}\left(\frac{d^2 - a^2 - b^2}{2ab}\right).$$

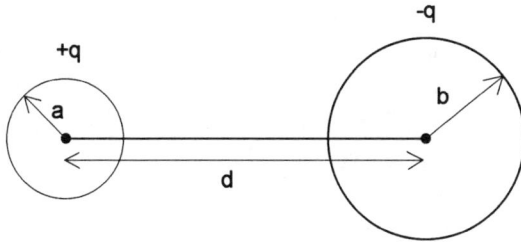

**Figure 2.25**    *Two spherical conductors of different radii.*

Often in electrostatic problems,though,we wish to know the capacitance between a sphere and an infinite flat plane as in Figure 2.26. This case is readily obtained from equation (2.49) by substituting a = R and b = (d - D) and allowing both d and b to become infinite. Following Smythe the solution is

$$C = 4\pi\varepsilon\varepsilon_0 R \sinh \alpha \sum_{n=1}^{\infty} \text{csch } n\alpha \quad (2.50)$$

where $\alpha = \cosh^{-1}(D/R)$. If D >> R,i.e. the distance of the sphere from the ground plane is large compared with its radius, then equation (2.50) gives the capacitance of an isolated sphere (cf. equation (2.48) with a = R ) as expected.

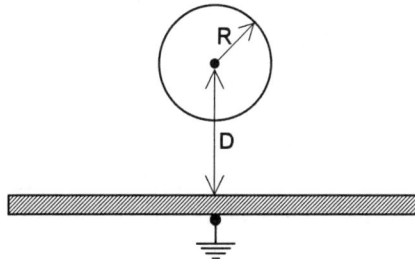

**Figure 2.26**    *Conducting sphere above an earthed surface.*

A spherical, 500-litre container, diameter about 1 m, located more than a few metres above ground, can be considered to be an isolated sphere. Its capacitance, therefore, will be about 56 pF. When it is closer to the ground and a high degree of accuracy is required, then equation (2.50) must be used.

### 2.7.3 Other Types of Capacitances

So far we have considered only the capacitance associated with ungrounded metal structures. It should not be forgotten that the human body behaves as a capacitor whose capacitance is about 150-200 pF. A charged plastic sheet located above and parallel to the flat base of a machine may be treated as a parallel-plate capacitor, in which case its capacitance will depend on the area of the sheet and its distance above the flat surface. Although conceptually more difficult to appreciate, any space charge, e.g. a charged dust cloud, has an associated capacitance.

### 2.7.4 Addition of Capacitances

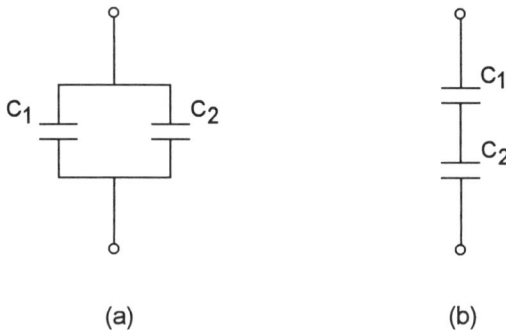

**Figure 2.27**    *(a) Parallel and (b) series combinations of capacitors.*

In each of the capacitances discussed in section 2.7.2, flux lines from the charged electrode flow through a single dielectric material and terminate on the same earthed surface. Therefore, a single capacitor element adequately describes the system. In an industrial situation, flux lines from charged product, or poorly earthed plant, may flow through two or more dielectric materials and may terminate on more than one earthed surface. The total capacitance associated with the charged body will then be a combination of two or more capacitors. Both parallel and series combinations may occur as in Figure 2.27(a) and (b).

*(a) Parallel Combination of Capacitors*

The total capacitance, C, of two capacitors connected in parallel is given by

$$C = C_1 + C_2. \tag{2.51}$$

An example of parallel capacitances is a charged plastic sheet resting on an earthed surface with a second earthed surface above (this could be a static measuring instrument perhaps) as in Figure 2.28(a). The charge on the upper surface of the plastic "sees" the two capacitances as being in parallel because the flux lines from the charge are shared between the two earthed surfaces.

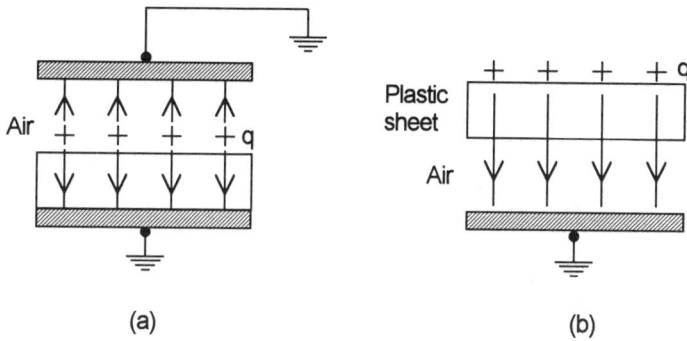

(a)                              (b)

**Figure 2.28**    *In (a) the flux from charge q on the dielectric surface is shared between two parallel capacitances while in (b) it passes through both dielectrics.*

*(b) Series Combination of Capacitors*
   Where capacitors are connected in series as in Figure 2.27(b), the total capacitance is given by

$$\frac{1}{C} = \frac{1}{C_1} + \frac{1}{C_2}. \tag{2.52}$$

Again a charged plastic sheet is taken as an example. This time the sheet is raised from the earthed surface on which it was resting and no other earthed surface is near. Now all the flux lines from the upper surface of the charged sheet in Figure 2.28(b) flow through both the plastic and the air but terminate on the same earthed surface. Conceptually we may represent both dielectrics by two capacitors in series as in Figure 2.27(b) and if, as is often the case, the air gap is very much larger than the film thickness then $C_2 \ll C_1$ and $C \sim C_2$.

**2.7.5  Energy Stored in a Capacitor**
   Irrespective of its shape or size, any capacitor across which a potential has developed is a store of electrical energy and it is the release of this stored energy

that can be so destructive in the electronics industry and can cause such catastrophic explosions and fires in the petrochemical, pharmaceutical and powder handling industries.

The amount of energy stored in a capacitor is readily deduced with the aid of Figure 2.29. In (a) a charge +q has already been transferred from the lower to the upper plate of a parallel-plate capacitor. The lower plate is connected to ground and its potential is zero. As a result of transferring the charge a potential, V, has

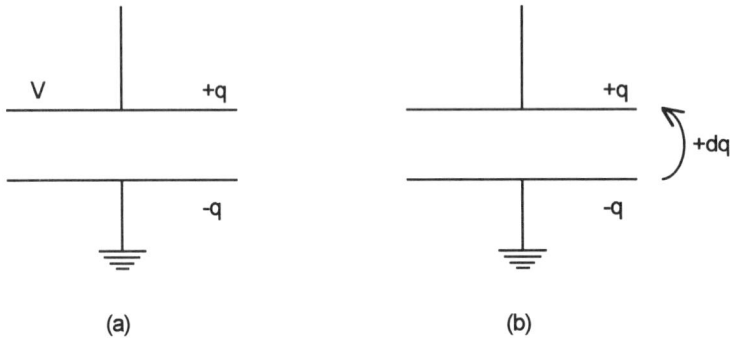

(a)                                    (b)

**Figure 2.29**    *Transfer of further charge between the plates of a charged capacitor requires work to be done against the electrostatic forces.*

developed on the upper plate. Now assume as in (b) that a further charge +dq is transferred to the upper plate. This charge is assumed to be sufficiently small that any change in the potential of the plate is insignificant. To effect this transfer, work must be done against the forces of repulsion of the positive charges already on the upper plate. From equation (2.20) the work done, dU, will be directly proportional to the potential of the plate and given by

$$dU = Vdq$$

which, from the definition of capacitance, may be written as

$$dU = \left(\frac{q}{C}\right)dq.$$

Of course, dU represents an increase in the potential energy of the charge. This energy is stored on the capacitor until the charge returns to the earthed plate, dissipating the energy as it does so. If we now commence with an uncharged capacitor and sum all the incremental changes in energy as each successive increment of charge is transferred to the upper plate, then after a total charge q has been transferred the total energy, U, stored on the capacitor is given by

$$U = \int_0^q \left(\frac{q}{C}\right) dq = \frac{1}{2}\frac{q^2}{C}. \tag{2.53}$$

An alternative form often used is obtained by substituting from equation (2.38) yielding

$$U = \frac{1}{2}CV^2. \tag{2.54}$$

Equations (2.53) and (2.54) are applicable to any capacitor, irrespective of the shape of the conductors. In this context, it should be remembered that personnel wearing insulating footwear and any metal pipes, containers or machinery where the earth bonding is poor are capacitors which can become stores of electrical energy if they become charged.

As an example, consider the section of metal pipe in Figure 2.23. If a hydrocarbon liquid or a nonconducting powder is flowing through it, this section of pipe could easily charge to a potential of 30 kV. Since the capacitance of the pipe is 16 pF, the total energy stored would be 7.2 mJ. We may use equation (2.54) also to calculate the electrical energy stored on operators. The body capacitance is about 200 pF and may readily charge to 5 kV corresponding to a stored energy of 2.5 mJ.

Many vapours and dusts have minimum ignition energies (MIE) in the range 1-10 mJ; for some very sensitive powders the MIE may be as low as a few $\mu$J. It is clear, therefore, that under the right conditions, the release of stored electrical energy from ungrounded conductors poses a real threat where materials with low MIE are present.

In the electronics industry even a few $\mu$J of energy is sufficient to cause catastrophic failure of sensitive devices. Therefore, ungrounded personnel can be particularly troublesome, because discharges of such small energy are well below the threshold of feeling in humans.

## 2.8 INDEPENDENT AND DEPENDENT ELECTROSTATIC PARAMETERS

In conventional electrical circuits, we are used to thinking of potential (voltage) as a well-defined quantity. Power supplies are generally designed to provide a controlled voltage output and the impedance of the circuit connected to the supply controls the current that flows. The voltage is then said to be the independent parameter, while the current is a dependent parameter.

In electrostatic systems, charge is usually the independent parameter. Consequently, potential becomes the dependent parameter and is generally ill-defined. As an example, consider the charged web passing over the roller in Figure 2.30(a). Assume that the web is 50 μm thick, has a relative permittivity of 3 and a charge density of 1μC/m² on its surface. The radius of the roller is sufficiently large compared with the thickness of the web for us to assume that the web behaves as a parallel-plate capacitor. This being the case, the potential at its free surface is easily calculated from equation (2.39) to be 1.9V.

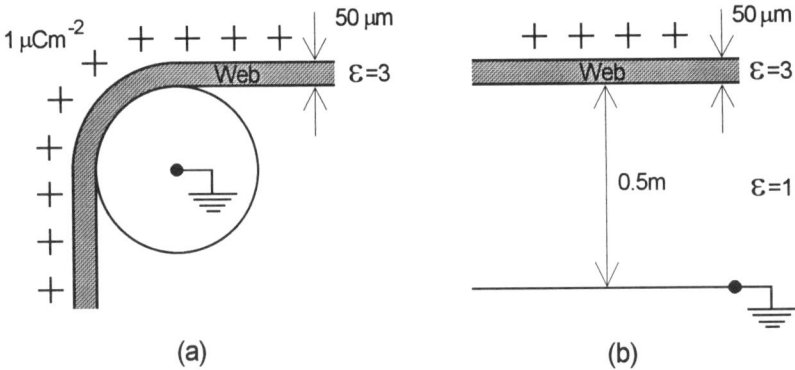

**Figure 2.30**    *(a) A charged insulating web passing over an earthed roller. (b) The same web suspended above the base of the machine.*

When the web moves well away from the roller (Figure 2.30(b)) so that the nearest earth is the base of the machine 0.5m below the web, say, the potential of the same surface will increase significantly. In this position, the web and the air gap may be treated as two capacitors in series so that the combined capacitance C is given by

$$C = \frac{C_{web}C_{air}}{C_{web} + C_{air}}.$$

Since $C_{web} \gg C_{air}$, it is readily seen that $C \sim C_{air}$. Thus the potential of the upper surface of the web will be essentially that of a parallel-plate air capacitor. This time the calculations yield a potential of 56.5 kV. This large difference in potential for two identically charged surfaces has arisen because of the dramatically different capacitances in the two cases. Not only has the distance of the charged surface from the nearest earthed plane changed, but the medium filling the space between the charged surface and earth has also changed.

A note of caution is necessary here for those setting out to monitor electrostatic effects in their process or on their product. The magnitude of the electrostatic charge is often quoted as a voltage read from the display of a static meter. It should be clear from the above argument that such a measurement could be misleading because as soon as the meter is brought near, the capacitance of the charge to earth changes and, therefore, so will its potential. When making measurements of static charge it is preferable to use a parameter such as electric field which is more readily converted into a charge density. However, even then caution is necessary because the measuring instrument itself, as well as other earthed surfaces, can distort the field.

## 2.9 INSULATING MATERIALS

Static-related problems are invariably associated with highly insulating materials such as plastic sheets, polymeric powders, man-made fibres and organic liquids. Insulators are at the high-resistivity end of a class of materials known collectively as dielectrics. They have a number of properties which are important in the context of static electricity. Some of these we have already met in earlier sections of this chapter, albeit implicitly, in the models that were developed. The intention here is to review these properties explicitly, thereby highlighting their relevance to electrostatics.

### 2.9.1 Relative Permittivity (Dielectric Constant)

In the preceding sections, we saw that the relative permittivity of a dielectric material determined the magnitude of the forces between electrostatic charges and, therefore, the electric field in the material. This is a consequence of the field-induced polarisation that occurs to a greater or lesser extent in all materials as they attempt to reduce their internal electric field. The relative permittivity is a measure of the extent to which they are successful in achieving this. Metals have infinitely high relative permittivies. The high conductivity resulting from a high concentration of free electrons ensures rapid and complete polarisation, the electric field in the metal falling to zero almost instantaneously. Insulators, on the other hand, have few if any free-charge carriers so their response to an electric field is limited and their relative permittivities are low. The response of insulating materials is more likely to arise from a small displacement of the charges in the constituent atoms and molecules (electronic/atomic polarisation) or from the rotation of permanent molecular dipoles (orientational polarisation).

The significance of the relative permittivity is perhaps best illustrated with reference to the two parallel-plate capacitors in Figure 2.31. The capacitors are identical except that one has a perfect vacuum between the plates while the second has a dielectric material between the plates. If we raise the potential of the vacuum capacitor to $+V$, a charge $+q_V$ must be transferrred to the upper plate, where $q_V = C_V V$.

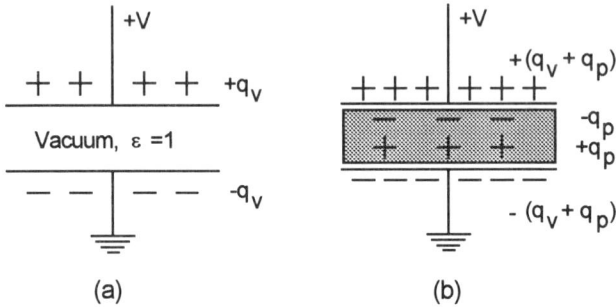

(a)                                    (b)

**Figure 2.31**    *Parallel-plate capacitors with (a) vacuum and (b) a dielectric material between the plates.*

When a polarisable material is present between the plates, then an additional charge $+q_p$ must be transferred to the upper plate to neutralise the effects of the negative charge appearing there as a result of the polarisation process. The capacitance in this case is now defined by

$$C_d = \frac{(q_V + q_P)}{V} = C_V \left(1 + \frac{q_P}{q_V}\right)$$

or

$$C_d = \varepsilon C_V = \frac{\varepsilon \varepsilon_0 A}{d}$$

where

$$\varepsilon = \left(1 + \frac{q_P}{q_V}\right).$$

Since $q_p$ and $q_V$ are both positive, $\varepsilon$ must be greater than unity, so $C_d > C_V$. It is seen, therefore, that polarisation in the material has increased the capacitance of the parallel-plate capacitor. The energy stored in the capacitor will also be higher, equation (2.54).

## 2.9.2 Volume Resistivity

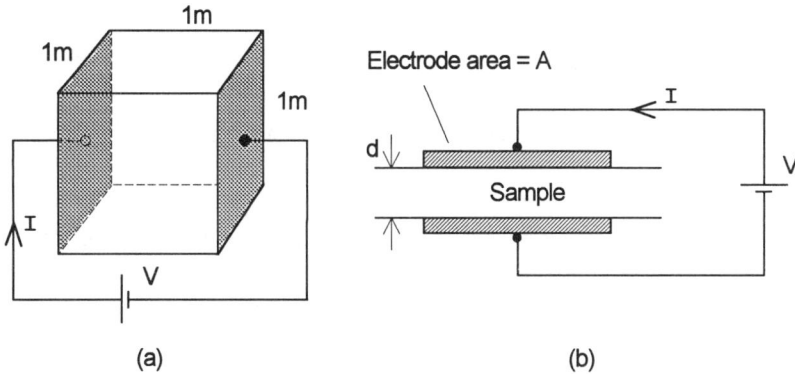

**Figure 2.32**    *The definition of resistivity is based on a cube of material of side 1 m as in (a).
To perform a measurement, samples in the form of sheets (b) are preferred.*

The volume (or bulk) resistivity of a material is defined as the resistance between opposite faces of a cube of the material of side 1 m   (Figure 2.32(a)). Obtaining and subsequently measuring the properties of such a large sample is not very practicable. Consequently, the resistivity is normally determined by measuring the resistance between opposite faces of a thin slab of the material as in Figure 2.32(b). The measured resistance, R, is then related to the resistivity, ρ, by the equation

$$R = \rho \frac{d}{A} \qquad (2.55)$$

where A and d are the electrode area and interelectrode spacing respectively. Conductors have low resistivity while insulators are characterised by a high resistivity.

The volume resistivity of a material is also the reciprocal of its conductivity, σ, which is defined by

$$\sigma = \sum_i n_i q_i \mu_i \qquad (2.56)$$

where $n_i$ and $\mu_i$ are the densities and mobilities of the free-charge carriers in the material and $q_i$ the charge on each carrier. Insulating materials possess very few free-charge carriers. The thermal generation of free electrons and ions in such materials is a very weak process. Charge carrier mobilities are also generally very low in these materials. Since both carrier concentrations and mobilities are small the conductivity will also be small. Typical values may be in the range $10^{-16}$ to $10^{-10}$ S/m.

### 2.9.3 Surface Resistivity

Film products and laminates are often characterised in terms of their surface resistivity. This parameter is measured using two electrodes of length L to define a square area on a flat surface of the material as in Figure 2.33. The surface resist-ivity is then defined as the resistance between these two electrodes. In principle, the size of the square is not important: increasing the length of the electrodes will decrease the measured resistance in the ratio 1/L while increasing the interelectr-ode gap increases the resistance by a factor L. However, measurements carried out at constant voltage may show a dependence on L because of the electric field dependence of the charge carrier generation and transport mechanism.

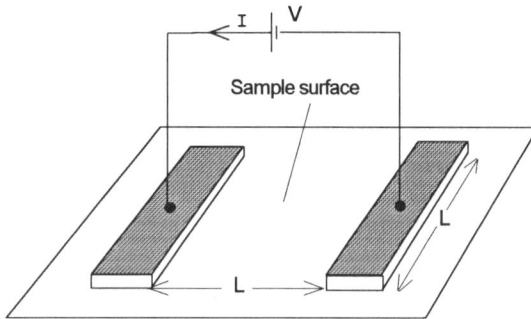

**Figure 2.33**    *Arrangement used for a surface-resistivity measurement.*

### 2.9.4 Relaxation Time

Under clean, dry conditions the surface resistivity of many plastic materials is greater than $10^{10}$ Ω per square and this coupled to their high volume resistivities means that any static charge generated in the insulator or on its surface cannot easily flow to earth. Indeed, an implicit assumption made throughout this chapter is that the charges are "static" and, therefore, totally unable to move.This is a valid assumption so long as the timescale for charge dissipation is long compared with the process time.

The charge dissipation time can be estimated in the following way. Assume that the parallel-plate capacitor, C, in Figure 2.34(a) is charged to some potential +V, by adding a charge +q to the upper plate. The voltage on this plate sets up an electric field in the insulator causing free charges to migrate. Negative charges migrate towards the upper plate and positive towards the lower plate. This charge

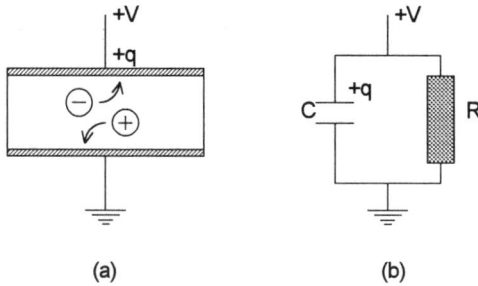

(a)                                              (b)

**Figure 2.34**   *(a) Decay of charge on a capacitor by conduction process in the dielectric and (b) the equivalent circuit for the process.*

flow constitutes a current through the insulator which results in a decrease in the charge on the upper plate. To represent correctly the capacitor in Figure 2.34(a) the equivalent circuit must be a parallel combination of an ideal capacitor (made from a perfect insulator) and a resistance, R, to account for the current flow through the insulator (Figure 2.34(b)). Those familiar with electrical circuits will know that when a capacitor, C, is charged to an initial voltage, $V_0$, and a resistance, R, connected across it, the voltage across the capacitor will decrease exponentially with time, t, according to the law

$$V = V_0 \exp{-\frac{t}{RC}}. \tag{2.57}$$

Using equation (2.38) the rate at which charge is lost from the capacitor, Figure 2.35, is given by

$$q = q_0 \exp{-\frac{t}{RC}} \tag{2.58}$$

From equations (2.55) and (2.40) we note that the product RC, the response time of the circuit, can be written as

$$RC = \rho \frac{d}{A} \times \varepsilon \varepsilon_0 \frac{A}{d} = \rho \varepsilon \varepsilon_0.$$

The analysis has shown, therefore, that the dissipation of charge through an insulator should follow an exponential decay law in which the time constant, $\tau$, for the decay is given by

$$\tau = \rho \varepsilon \varepsilon_0. \tag{2.59}$$

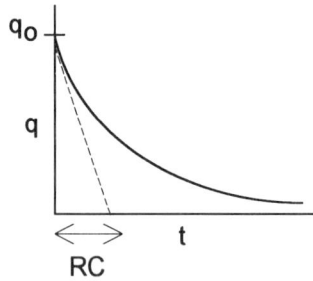

**Figure 2.35**    *Time-dependence of the charge on a capacitor shunted by a resistor. When the decay is exponential extrapolation of the initial slope to the time axis yields the time constant for the decay.*

The time constant, $\tau$, is known as the relaxation time of the insulator and is independent of the geometry of the insulator. In an insulator for which $\rho = 10^{12}$ $\Omega$m and $\varepsilon = 2.5$, the calculated relaxation time is

$$\tau = 10^{12} \times 2.5 \times 8.85 \times 10^{-12} = 22.1 \;\; s.$$

The importance of the relaxation time in electrostatics can be seen in the following examples. Consider a polymer film travelling at 10 m/s between earthed rollers spaced 2 m apart. The time taken for a particular section of the sheet to travel this distance is 0.2s. If the charge relaxation time of the polymer exceeds a few seconds then very little dissipation will occur from the film during transit between the rollers.

Contrast this with a filling operation in which an insulating liquid is being piped into a large storage vessel. Assume that the total volume of the vessel is 500 m³ and that it is being filled at the rate of 50 L/s. The total time to fill the container will be $10^4$s, i.e. just under 3 hours. If the charge relaxation time in the liquid is about 20s then at this low flow rate most of the charge is dissipated during the filling operation and cannot accumulate to dangerous levels.

These examples highlight two methods for controlling static in industry:

(i) Use additives where possible to increase the conductivity of the material

(ii) Keep the velocity of the material low.

Unfortunately, these methods are not always compatible with production goals!

When $\rho < 10^8$ $\Omega$m,i.e. $\tau < 10^{-3}$s,then no electrostatic problems are likely to be encountered in a production system. If $\rho > 10^8\Omega$m then static may well accumulate in the system, the degree of accumulation depending on several factors including the speed of the process. It should be obvious from the above discussion and the examples given that, as production speeds increase, static effects will become apparent in materials of higher conductivity (lower resistivity), materials which hitherto may not have shown any such effects.

The analysis leading to equations (2.58) and (2.59) suggests that the decay of static charge in a material should be exponential with a single relaxation time equal to $\tau$. In practice, this is not the case. The relaxation time is only given by equation (2.59) if all the flux from the charge is contained within the same insulator at all times. For the charge residing on the surface of the plastic sheet in Figure 2.28(a), for example, the time constant will be modified by the capacitance of the air gap. Furthermore, in real insulators, the volume and surface resistivities may not be well-defined (see Chapter 7). This can result in the charge decay characteristics of nonconductors departing significantly from the ideal exponential behaviour expected.

### 2.9.5 Transparency to Electric Flux

All insulating materials are "transparent" to electric flux lines. This property has an important bearing on the measurement of electrostatic charges. Consider the insulating plastic sheet in Figure 2.36 which has charges on both its surfaces. When a static meter is brought near to the sheet electric field lines from charges on both surfaces will terminate on the meter. Thus, without further experimentation it is not possible to distinguish one set of charges from the other.

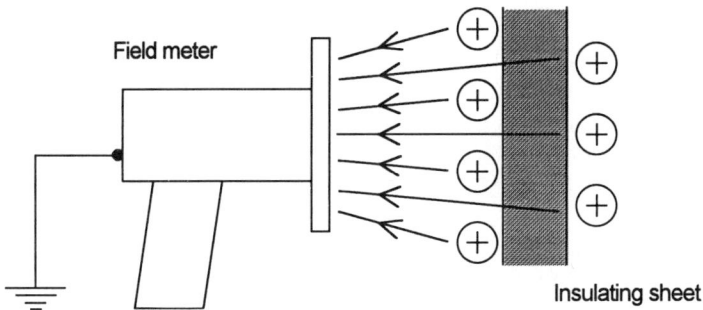

**Figure 2.36**    *A fieldmeter will measure charge from both surfaces of an insulating sheet.*

### 2.9.6 Breakdown Strength

As discussed above, insulators have very few free-charge carriers so,even when subjected to large electric fields, very little current flows through them. However, every material has a critical electrical field above which its insulating properties fail catastrophically. This critical electric field is termed the breakdown strength of the material. (Some texts also refer to dielectric strength. This term is not used in this book because of possible confusion with dielectric constant). The breakdown process is characterised by a high current flow through bright localised channels (the spark channel).

Under normal atmospheric conditions air has a breakdown strength of 2.7 MV/m, and most gases have breakdown strengths in the range 2.5 - 10 MV/m under a pressure of 1 atmosphere (Meek and Craggs, 1978). Simple hydrocarbon liquids have breakdown strengths in the range 140 to 200 MV/m (Gallagher, 1975). For polymeric materials much higher breakdown strengths have been reported with values ranging from 500 to 800 MV/m (O'Dwyer, 1973).

## 2.10  EQUIVALENT CIRCUIT OF AN ELECTROSTATIC SYSTEM

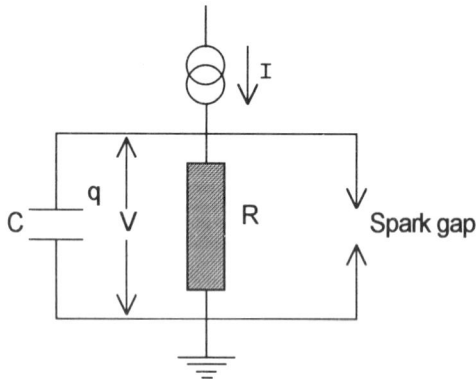

**Figure 2.37**    *An equivalent circuit for describing an electrostatic system. See text for the origin of the various components.*

An electrostatic system may be represented by the simple electrical equivalent circuit in Figure 2.37. It consists of four elements only. The first is a current generator, I, to represent the charge generating mechanism; in a typical production operation,charge generation will be a continuous process. As we shall see later, the current can be generated during pipeline flow, during a sieving operation or during winding operations involving insulating webs or films. The second element is the capacitor, C, on which charge is stored. The capacitor may be the surface of an insulating roller, or perhaps some metalwork or personnel not properly earthed. The third element is the resistance, R, which represents the charge relaxation mechanism in the electrically stressed insulator. The relaxation process will usually be a small current flow either through the bulk of the insulator or along its surface. Finally, we include a spark gap which limits the maximum voltage that can be attained in the system.

At the start of a production run, we may assume that the capacitor is uncharged. The current from the generator will initially charge up the capacitor causing the voltage, V, across it to increase. As the voltage rises, some of the current will begin to flow through the resistance, R. Further increases in V result in a higher and higher proportion of the generated current flowing through R. In the absence of other effects, the system eventually comes into equilibrium when all the generated current leaks to earth through R.

The instantaneous voltage across the capacitor is given by Ohm's law and reaches an equilibrium value, $V_e$ given by

$$V_e = IR. \tag{2.60}$$

Whether the system gives rise to significant electrostatic effects depends on the magnitude of $V_e$ and in particular on the magnitude of the resulting electric field in the insulator across which this voltage appears.

As an example, consider an industrial process which generates electrostatically a current of $10^{-9}$ A. If the resistance of the system to earth is $10^{14}\,\Omega$ (quite possible with today's insulators), then a voltage of 100 kV could develop in the system. If the electric field associated with this voltage is sufficient to cause electrical failure of the surrounding insulator (usually air) then a spark will occur. This effect is represented by the spark gap in the equivalent circuit.

Assuming that the system capacitance is 10 pF and that breakdown occurs when the system potential reaches 100 kV, the energy released in the discharge would be 50 mJ. This is more than sufficient to cause the ignition of some dust clouds and organic vapours and is also sufficient to cause discomfort to personnel.

Once the capacitor has discharged, its voltage falls to zero and the whole cycle starts afresh with the capacitor charging up to the breakdown condition again.

The role of environmental factors in governing the incidence of static problems can be understood when it is realised that changes in humidity or in surface cleanliness of insulators can change the value of the resistance R by many orders of magnitude. The model shows also that by increasing production speeds, i.e. increasing I, the equilibrium voltage for the system is increased. Thus a process that was previously free of any electrostatic problem may develop one as production rates increase.

## REFERENCES

Cross JA 1987 *"Electrostatics: Principles, Problems and Applications"* (Bristol: Adam Hilger).
Gallagher TJ 1975 *"Simple Dielectric Liquids: Mobility, Conduction, and Breakdown"* (Oxford:Clarendon).
Horvath T and Berta I 1975 "Mathematical simulation of electrostatic hazards" *IOP Conf Ser No 27* 256-263.
McAllister D, Smith JR and Diserens NJ 1985 *"Computer Modelling in Electrostatics"* (Letchworth: Research Studies Press and New York: Wiley).
Meek JM and Craggs JD, eds 1978 *"Electrical Breakdown of Gases"* (Chichester:Wiley).
Nayfeh MN and Brussel MK 1985 *"Electricity and Magnetism"* (New York: Wiley).
O'Dwyer JJ 1973 *"The Theory of Electrical Conduction and Breakdown in Solid Dielectrics"* (Oxford: Clarendon).
Smythe WR 1968 *"Static and Dynamic Electricity"* 3rd Ed (New York: McGraw-Hill).

*CHAPTER THREE*

# INITIATION OF ELECTROSTATIC PHENOMENA

In Chapter 1 we saw that for many industrial processes the generation of static electricity is essential for effecting a particular process step. We saw also that a variety of static-related problems can arise in a wide range of different industries. In view of these many and diverse ways in which static can manifest itself, it is perhaps surprising to discover that relatively few physical mechanisms lead to all the effects that are observed.

Most electrostatic problems in industry occur when unwanted, uncontrolled charging occurs and can usually be traced to the fact that somewhere in the process the product, whether liquid or solid, came into contact with other materials. As we will see later, separation of positive and negative charges readily occurs at the interface between dissimilar materials. Consequently, one of the materials becomes positively charged while the other becomes negatively charged. Whether this process leads to a significant static problem depends on how rapidly the separated charges can dissipate once the contact is broken. When one, or both, of the contacting materials is a good electrical insulator then dissipation is slow and static charges will accumulate either on or within the insulator.

Industries that rely on static charge to achieve a particular process step, require controlled sources of static charge and generally make use of corona charging or induction charging. Such sources give reliable operation over long periods of time and,when problems do arise, these may often be traced to changing environmental factors or equipment degradation.

In this chapter the mechanisms that lead to the generation of static charge will be described in some detail and examples given of relevant industrial situations in which particular mechanisms operate.

## 3.1 CHARGING OF LIQUIDS IN PIPELINE FLOW

It has been known for many years that petrochemicals become highly charged unless stringent precautions are taken during the handling and transportation of such liquids. (An excellent introductory review is the book by Klinkenberg and Van der Minne (1958)). The charging phenomena which occur in these insulating hydrocarbon liquids are amongst the most widely studied and understood of all electrostatic charging mechanisms. Clearly, the prime motivation has been to reduce the risk of electrostatically-induced explosions when handling such flammable liquids. Of particular concern to the petrochemical industry is the charging which occurs during pipeline flow. The origin of the effect is the formation of a double-layer of charge at the interface between the liquid and the wall of the pipe. During flow, this double-layer is sheared and charge becomes entrained in the liquid. If this charge is allowed to accumulate in the system a potentially hazardous situation is created.

### 3.1.1 The Double-Layer

It was Helmholtz in 1879 who first postulated that a double-layer invariably formed at a phase boundary. From this basic concept he developed a mathematical theory relating the velocity of electroosmotic flow to the charge separation which occurred in the double-layer. Despite this early success, some thirty years elapsed before Gouy (1910) in France and Chapman (1913) in Britain independently presented a theory which described the structure of the double-layer. The so-called Gouy-Chapman theory, which was applied initially to the interface between solids and weak electrolytes has since been added to by Stern (1924) and by Grahame (1950). In recent years yet further refinement has taken place (Henderson and Blum, 1978; Outhwaite and Bhuiyan, 1983; Torrie et al, 1982).

Fortunately, most of these later refinements apply only to strong electrolytes. Most organic liquids are highly insulating, with relatively few free-charge carriers. They can, therefore, be thought of as weak electrolytes since no matter how well such liquids are refined and purified, they will inevitably contain minute quantities of ionisable impurities. It is the dissociation of impurities into positive and negative ions that determines the low-field conductivity of these liquids. When they are moderately pure, the conductivity of most organic liquids is in the range 0.1 to 10 pS/m. Obviously, the more a liquid is handled after purification the more likely it is to become contaminated, in which case its conductivity will increase.

Ions produced by dissociation of trace contaminants undergo random thermal motion which distributes them uniformly throughout the liquid. This situation changes at a boundary with a solid surface because one type of ion, say positive, is preferentially adsorbed to the surface of the solid (Figure 3.1). If the liquid was electrically neutral at the beginning, the loss of positive charge to the solid surface means that an excess negative charge of equal magnitude must be present in the liquid. Thus, a difference in potential is created between the solid and the bulk liquid. With positive ions preferentially adsorbed to the surface, as in Figure 3.1, the potential of the solid surface must be positive relative to the bulk liquid. The magnitude of this potential depends on the density of the ions localised at the solid surface.

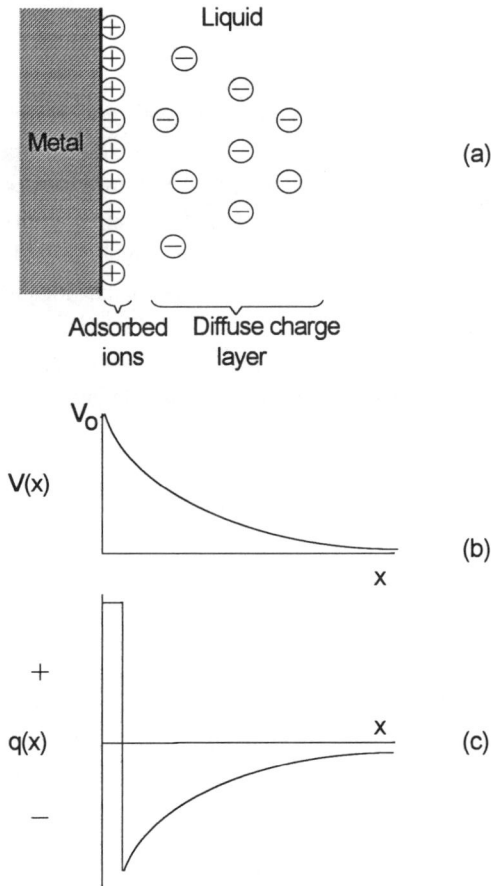

**Figure 3.1**    *(a) A double-layer at a solid-liquid interface. (b) The potential and (c) the charge distribution across the interface.*

The separation of positive and negative charges at the interface gives rise to an electric field in the liquid which causes the ions in the liquid to redistribute; negative ions are attracted towards the surface, positive ions are repelled. Consequently, the concentration of negative ions near the interface is increased above that in the bulk liquid while the positive-ion concentration is reduced. Diffusion processes now become important. As a result of the concentration gradients set up near the interface, negative ions diffuse away from the interface, while positive ions diffuse towards the interface. This diffusive flow of ions opposes the redistribution (conduction) caused by the electric field originating from the charged surface. Eventually, the diffusive and electrical flows balance and the interface comes into equilibrium.

The immobile surface charge and the diffuse layer of counter ions in the liquid (see Figure 3.1) together constitute the double-layer.

In the Gouy-Chapman model of the double-layer, it is assumed that the adsorbed surface charge is distributed uniformly over the surface of the solid and that the counter ions in the diffuse part of the double-layer are point charges. The ions in the liquid must satisfy simultaneously both the Poisson equation for space-charge (equation 2.23) and the Boltzmann equation which provides a mathematical relation between the concentration of ions at a particular location in the liquid and the electrostatic potential there.

Applying the Boltzmann equation to the case in hand enables the concentrations of positive and negative ions, $n(x)^+$ and $n(x)^-$ respectively, at a distance x from the solid surface in Figure 3.1(a), to be written as

$$n(x)^+ = n \exp\left(-\frac{eV(x)}{kT}\right) \qquad (3.1)$$

$$n(x)^- = n \exp\left(+\frac{eV(x)}{kT}\right) \qquad (3.2)$$

where $n = n(\infty)^+ = n(\infty)^-$ is the density of ion pairs in the bulk liquid well away from the interface, e the electronic charge, k Boltzmann's constant and T the absolute temperature. (Here we have assumed the ions to be monovalent). The charge density, q(x), at x is then given by

$$q(x) = e[n(x)^+ - n(x)^-]$$

i.e.

$$q(x) = -en\left[\exp\left(\frac{eV(x)}{kT}\right) - \exp\left(-\frac{eV(x)}{kT}\right)\right] \qquad (3.3)$$

which on substituting into Poisson's equation yields

$$\frac{d^2V(x)}{dx^2} = \frac{ne}{\varepsilon\varepsilon_o}\left[\exp\left(\frac{eV(x)}{kT}\right) - \exp\left(-\frac{eV(x)}{kT}\right)\right]. \qquad (3.4)$$

For large values of x, the electric field of the charged surface, i.e. the potential gradient in the liquid, must decrease to zero (screened by the diffuse cloud of counter ions). Furthermore, if we assume that the potential in the bulk liquid is also zero and that insulating liquids may be considered to be weak electrolytes so that $eV(x) \ll kT$, then equation (3.4) may be integrated twice (Sennet and Olivier, 1966), yielding

$$V(x) = V_0\exp\left(-\frac{x}{\delta}\right) \qquad (3.5)$$

where $V_0$ is the potential at the surface of the solid. $\delta$, which has the dimensions of length, is given by

$$\delta = \left(\frac{\varepsilon\varepsilon_o kT}{2ne^2}\right)^{\frac{1}{2}}. \qquad (3.6)$$

According to the Einstein relation, the ratio of the mobility, $\mu$, to the diffusion coefficient, D, of ions is given by

$$\frac{\mu}{D} = \frac{e}{kT}$$

which on substituting into equation (3.6) yields

$$\delta = \left(\frac{\varepsilon\varepsilon_o D}{2ne\mu}\right)^{\frac{1}{2}}.$$

For a monovalent electrolyte in which both species of ion have identical mobilities, this can be further simplified using equations (2.56) and (2.59) to give

$$\delta = (D\tau)^{\frac{1}{2}} \qquad (3.7)$$

where $\tau$ is the dielectric relaxation time of the electrolyte.

Equation (3.5) shows that the potential in the liquid decreases exponentially from an initial value $V_0$ at the interface to zero in the bulk liquid (Figure 3.1(b)). If we now substitute for V(x) in equation (3.3) while retaining the assumption of a weak electrolyte, we obtain the charge density in the diffuse part of the double-layer as

$$q(x) = -\left(\frac{2ne^2 V_o}{kT}\right) \exp\left(-\frac{x}{\delta}\right) \tag{3.8}$$

which can be written as

$$q(x) = -\left(\frac{\varepsilon\varepsilon_o V_o}{\delta^2}\right) \exp\left(-\frac{x}{\delta}\right) . \tag{3.9}$$

We see, therefore, that the charge density in the liquid (Figure 3.1(c)) also decreases exponentially into the bulk of the liquid in the same way as the potential.

Since the system as a whole must remain electrically neutral, the net negative charge in the diffuse layer must equal the net positive charge on the solid surface. Assuming, therefore, that the liquid extends to infinity for positive values of x but at x = 0 is bounded by a solid surface of area, A, to which a uniform charge +Q has adsorbed, we may write

$$Q = -\int_0^\infty Aq(x)dx = \int_0^\infty A\left(\frac{\varepsilon\varepsilon_0 V_0}{\delta^2}\right) \exp\left(-\frac{x}{\delta}\right) dx \tag{3.10}$$

which simplifies to

$$Q = \left(\frac{\varepsilon\varepsilon_0 A}{\delta}\right) V_0 . \tag{3.11}$$

This is a particularly interesting result since it shows that a double-layer behaves exactly like a parallel-plate capacitor in which the plate separation is $\delta$ and across which a potential V appears (see equation 2.39). Until now, the parameter $\delta$ has been considered simply as a characteristic length in the exponential decay of charge and potential through the double-layer. By analogy with the charged capacitor, we see now that it is a measure of the effective width of the double-layer. Table 3.1 gives some calculated values for the double-layer width for various ionic concentrations. It is seen that as ionic strength decreases (lower conductivity) so the double-layer extends further and further out from the solid surface.

The total charge in the diffuse layer depends on the double-layer potential, $V_0$, which usually has values in the range 10-500mV (and can be positive or negative) depending on the nature of both the solid material and the ions in the liquid.

From the preceding discussion we see that the formation of a double-layer at a phase boundary is a naturally occurring phenomenon. So far, though, we have assumed implicitly that the liquid adjacent to the surface is stationary. When the ionic strength, n, is small - usually the case for organic liquids - the double-layer is sufficiently extensive that it becomes subjected to hydrodynamic forces, i.e. convection in the liquid. It is under these conditions that static related problems arise. When the liquid begins to flow the diffuse part of the double-layer shears, thus effecting a separation of the positive and negative charges forming the double-layer.

**Table 3.1.**   *Double-layer widths calculated from equation (3.6) for different ionic strengths and assuming $T = 300 \, K$ and $\varepsilon = 2$. Also included are the corresponding conductivities assuming $\mu = 1x10^{-7} \, m^2V^{-1}s^{-1}$ (Gallagher, 1975).*

| n (m$^{-3}$) | $1x10^{13}$ | $1x10^{14}$ | $1x10^{15}$ | $1x10^{16}$ |
|---|---|---|---|---|
| $\delta$ ($\mu$m) | 378 | 120 | 38 | 12 |
| $\sigma$ (pS/m) | 0.16 | 1.6 | 16 | 160 |

### 3.1.2 The Streaming Current

When a liquid flows through an earthed, metal pipe, a double-layer forms at the inner wall of the pipe as described above. Now, however, in addition to electrical conduction and diffusion effects, the ions in the diffuse layer are subjected to convective flows, since they are entrained in the liquid. The most obvious result is that the diffuse part of the double-layer is swept out of the pipe, as shown in Figure 3.2, giving rise to an electrical current, $I_s$, known as the streaming current. This current is easily measured by collecting the liquid in a Faraday pail (section 6.1). For every negative ion leaving the pipe, a positive ion is discharged at the pipe wall. This current to the pipe wall continues so long as liquid flow is maintained through the pipe. The current is equal in magnitude but opposite in sign to that which flows to the Faraday pail and can be measured by connecting the pipe to earth through an appropriate meter as shown in Figure 3.2.

In principle, calculating the magnitude of the streaming current, $I_s$, is relatively easy. We need only know the charge density, $q(r)$, in the liquid at the pipe outlet and the velocity profile $v(r)$ of the emerging liquid stream. The streaming current is then obtained by integrating the product of charge density and flow velocity over the pipe cross-section, i.e.

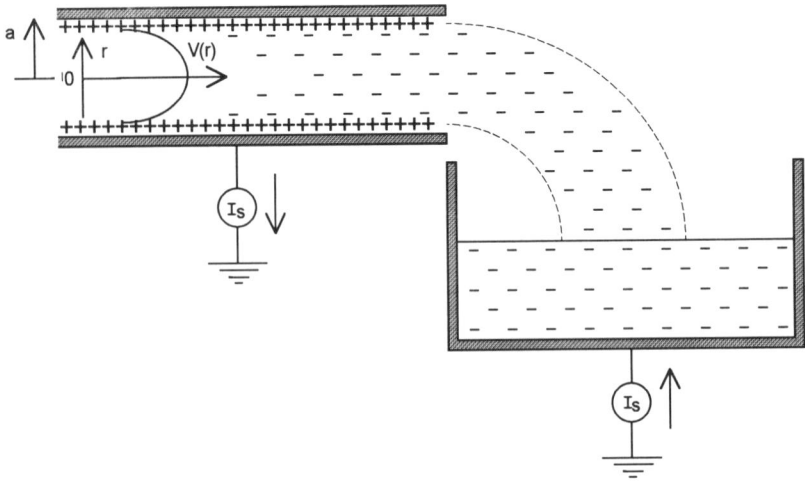

**Figure 3.2**    *Measuring the streaming currrent generated when a liquid flows through a conducting pipe.*

$$I_S = 2\pi \int_0^a rq(r)\mathrm{v}(r)dr \ . \tag{3.12}$$

### 3.1.2.1 Streaming Current in Long Pipes

The earliest theories (Klinkenberg and Van der Minne,1958; Rutgers et al, 1957) assumed that the pipe through which the liquid flowed was sufficiently long to establish equilibrium conditions in which case the charge in the diffuse part of the double-layer is given by equation (3.9). The velocity profile of a liquid undergoing laminar flow in a pipe is given by

$$\mathrm{v}(r) = 2\mathrm{v}_{av}\left(1 - \frac{r^2}{a^2}\right) \tag{3.13}$$

where a is the pipe radius and $\mathrm{v}_{av}$ the average velocity at which liquid emerges from the pipe. Assuming that the double-layer thickness is much smaller than the pipe radius, the diffuse part of the double-layer is confined to a thin layer close to the wall where the velocity profile is linear and given by $(4\mathrm{v}_{av}x/a)$ where x is the distance from the pipe wall. Thus, the streaming current, $I_\infty$, generated by a liquid in laminar flow in an infinitely long pipe is given by

$$I_\infty = -2\pi a \int_0^\infty (4v_{av}x/a)(\varepsilon\varepsilon_0 V_0/\delta^2)e \exp(-x/\delta)dx. \qquad (3.14)$$

Strictly, the lower limit of integration should be the shear plane which lies a few molecular diameters from the pipe wall. At this plane the potential in the liquid is known as the zeta-potential (Sennet and Olivier, 1966) and is given the symbol $\zeta$. The upper limit of the integral in equation (3.14) should be small compared with the pipe radius in order to use the linear approximation for the velocity profile near the wall. However, because the double-layer is so thin, the exponential term under the integral falls so rapidly with increasing x that an upper limit equal to infinity is permissible and introduces very little error. Carrying out the integration subject to these conditions yields

$$I_\infty = -8\pi\varepsilon\varepsilon_0\zeta v_{av} \qquad (3.15)$$

for the streaming current. When the double-layer thickness becomes comparable with the pipe radius, as would be the case with liquids of low conductivity, Taylor et al (1974) have shown that the streaming current in laminar flow is given by

$$I_\infty = -\frac{8\pi\varepsilon\varepsilon_0 v_{av}}{I_1(a/\delta)}\frac{\Gamma(2)}{\Gamma(0)} \sum_{k=0}^\infty (-1)^{k+1}(2k+2)\frac{\Gamma(k)}{\Gamma(3+k)} I_{2k+2}\left(\frac{a}{\delta}\right) \qquad (3.16)$$

where $I_n(a/\delta)$ are Bessel functions of order n and $\Gamma$ is the Gamma function. Equation (3.16) reduces to equation (3.15) when $a>>\delta$. From the above analysis it is seen that the magnitude of the streaming current is proportional to the average flow velocity so long as the liquid remains in laminar flow. When the flow becomes turbulent, thorough mixing of the liquid in the turbulent core occurs. Therefore, the ions in that part of the diffuse double-layer which lies outside the laminar sublayer become uniformly dispersed throughout the pipe. The streaming current is then given by

$$I_\infty = \pi a^2 q_{av} v_{av}$$

where $q_{av}$ is the average charge density in the turbulent core. Assuming that $\delta$ greatly exceeds the width of the laminar sublayer, and since charge neutrality must hold in any section of the pipe, then from equation (3.11) the average charge density is deduced to be $(2\varepsilon\varepsilon_0\zeta/a\delta)$ so that the streaming current generated during turbulent flow in a long pipe is given by

$$I_\infty = -2\pi(a/\delta)\varepsilon\varepsilon_0 \zeta v_{av}. \qquad (3.17)$$

Thus
$$\frac{I_\infty(turb)}{I_\infty(lam)} = \frac{a}{4\delta}$$

from which we deduce that at the transition velocity from laminar to turbulent flow the streaming current is expected to increase suddenly by a factor $a/4\delta$. This effect was demonstrated experimentally by Rutgers et al (1957) who showed that ,when turbulence set in, the current could increase by as much as 60 times that in laminar flow because $\delta$ is much smaller than the pipe radius (see Table 3.1).

### 3.1.2.2 Effect of Flow Velocity

Equation (3.17) suggests that the streaming current in turbulent flow should increase linearly with flow velocity. In practice, the dependence is found to be superlinear, often close to a square law. According to Gavis and Koszman (1961) the earlier models based on the classical double-layer theory were unsatisfactory because the detailed behaviour of ions near the pipe wall was neglected. They argued that the streaming current was limited by a form of concentration polaris-ation at the pipe wall. Thus the rate at which positive ions can discharge at the wall in Figure 3.2 is limited by diffusion across a thin layer adjacent to the wall. Solving the relevant equations (Gavis and Koszman, 1961; Koszman and Gavis, 1962a) leads to

$$I_\infty = K d^{0.88} v_{av}^{1.88} \qquad (3.18)$$

in which d is the pipe diameter and K is given by

$$K = (0.035 \varepsilon\varepsilon_0 kT/ne v^{0.63} D^{0.25})(1 - C_S/C_0) \qquad (3.19)$$

where n is the transference number of the non-adsorbing ion, $v$ the kinematic viscosity of the liquid, $C_S$ and $C_0$ the concentrations of the adsorbing ion at the pipe surface and in the turbulent core respectively. Experiments with small dia-meter pipes (Koszman and Gavis,1962b; Gibson and Lloyd, 1970a) gave good agreement with theory.

Following a series of measurements on the electrification of toluene flowing in large diameter pipes Gibson and Lloyd (1970b) concluded that the dependences on flow velocity and pipe diameter were different to those found in small pipes, the streaming current now following an empirical law of the form

$$I_\infty = K_1 d^n \mathrm{v}_{av}^m \qquad (3.20)$$

in which n~1.6, m ~ 2.4 and $K_1$ a constant in the range $5 \times 10^{-7}$ to $7 \times 10^{-6}$ if d is in metres and $\mathrm{v}_{av}$ in m/s. Their results confirmed earlier measurements by Schon (1965) who found that the electrification of "motor spirit" flowing through long pipes of various diameters in the range 2.5 to 20 cm followed a similar law, namely

$$I_\infty = K_2 d^\alpha \mathrm{v}_{av}^\alpha \qquad (3.21)$$

where $\alpha$ had a value between 1.8 and 2.0 and $K_2 = 4 \times 10^{-6}$. Both equations give similar results and predict that the streaming currents from long pipes are in the range $10^{-10}$ to $10^{-7}$ A.

Since the streaming current is found generally to depend only weakly on surface or solute materials as predicted by their theory Gavis and Koszman concluded that the electrification process must be limited by the rate at which the adsorbing species diffuses to the pipe wall. However, this is disputed by Walmsley and Woodford (1981) and Walmsley (1982) who suggest that as liquid conductivity increases, the *kinetics* of the adsorption process become important. In a general theory which includes the adsorption and diffusion rates for both the positive and negative ions they show that several different regimes of behaviour exist depending on the conductivity of the liquid. In general, for low conductivity liquids, electrification is diffusion limited as predicted by Gavis and Koszman,but at higher conductivities adsorption becomes the limiting process. This finding is important because it shows that the polarity of the streaming current may change when additives are introduced into the liquid in order to increase its conductivity.

### 3.1.2.3 Effect of Pipe Length

In the above, we have assumed that the pipe through which the liquid is flowing is sufficiently long to enable the formation of an equilibrium double-layer. When flow begins, however, fresh uncharged liquid enters the pipe. During its passage through the pipe this liquid gradually loses positive ions, which diffuse to the pipe wall where they are preferentially adsorbed[1]. This results in the accumulation of excess negative charge in the liquid. The electric field between the adsorbed positive charge and the negative charge in the liquid causes a conduction current to flow in opposition to the diffusion current. When the pipe is sufficiently long these

---

[1]     *For some liquid/pipe combinations, negative ions may preferentially adsorb to the pipe wall.*

processes come into equilibrium and a fully formed double-layer is established at the interface.

From this qualitative argument, we can deduce that the streaming current from a short pipe will be small but that it increases to some maximum value as the pipe length increases.

To determine the dependence explicitly, it is necessary to consider simultaneously the effects of conduction, diffusion and convection on the charge density in the liquid. The general equation of continuity, which describes the charge density $q(r,x)$ in a flowing liquid at a radial position r and at an axial distance x from the pipe inlet, is given by

$$v(r)\frac{\partial q(r,x)}{\partial x} - \frac{D}{r}\frac{\partial}{\partial r}\left(r\frac{\partial q(r,x)}{\partial r}\right) + \frac{q(r,x)}{\tau} = 0. \qquad (3.22)$$

Equation (3.22) was solved by Gavis and Koszman (1961) for the case of turbulent flow yielding

$$I_S = I_\infty\left(1 - \exp\left(-\frac{L}{v_{av}\tau}\right)\right) \qquad (3.23)$$

where L is the length of the pipe and $I_\infty$ is given by equation (3.18). We see from their solution that as the residence time, $L/v_{av}$, of the liquid in the pipe increases the streaming current increases to the value expected for an infinitely long pipe, this occurring when the residence time exceeds the dielectric relaxation time, $\tau$, of the liquid.

For liquids in laminar flow it was shown by Taylor et al (1974) that equation (3.23) still applied except now $I_\infty$ is given by equation (3.16) and the average flow velocity is replaced by an effective velocity, $v_{av}/\beta$ where $0.5 < \beta < 1$.

### 3.1.2.4 Effect of Liquid Conductivity

Conductivity has a major influence on the degree of electrification of a liquid in pipeline flow as can be seen in Figure 3.3. When the conductivity is low, usually associated with pure liquids, the streaming current is low. As the conductivity increases, the current increases rapidly. The arguments given above for the effect of pipe length may be used again here to explain this behaviour. Increasing the conductivity of the liquid decreases its relaxation time so that equilibrium is attained for much shorter residence times in the pipe. Thus, for a given length of pipe, $I_S$ will increase as $\tau$ decreases (equation (3.23)). However, on increasing the conductivity further, the current is expected to saturate at a value corresponding to

$I_\infty$ . Clearly, this is not the case. The current passes through a maximum and then begins to decrease.

Qualitative arguments have been presented which suggest that such behaviour can be expected when the double-layer becomes thinner than the laminar sublayer. However, the detailed model presented by Walmsley (1982) shows that,while processes within the pipe can give rise to this effect, the maximum is too broad and the calculated fall in current too small to explain the experimental results. He suggests an effect of pipe roughness as the cause of the discrepancy but the most likely explanation is that the streaming current is subjected to outlet effects as described below.

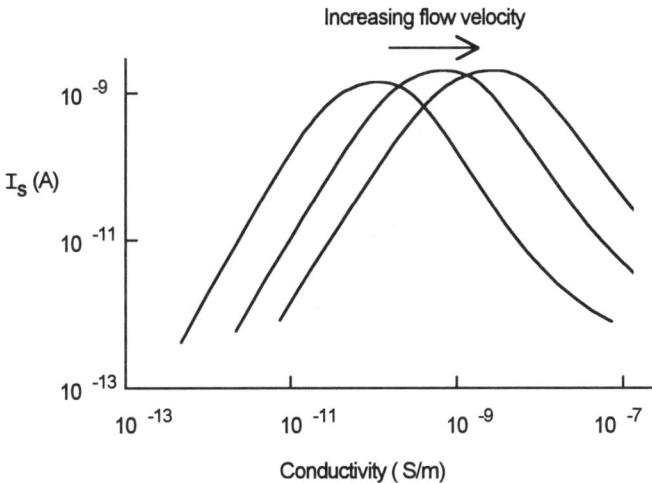

**Figure 3.3** *Variation of streaming current with liquid conductivity for different liquid flow rates.*

### 3.1.2.5 Outlet Effects

In section 2.9.4 we saw that charges in or on insulators dissipate to earth with a time constant given by the relaxation time of the insulator. Charges in the liquid stream emerging from the earthed pipe in Figure 3.4 will also try to relax to the nearest earth; in this case the pipe. Taylor (1974) showed that the charge relaxation effect could be accounted for by including an extra term in the equations for streaming current. Thus the Gavis and Koszman equation for turbulent flow should be written as

$$I_S = I_\infty \left(1 - \exp\left(-\frac{L}{v_{av}\tau}\right)\right) \exp\left(-\frac{x_0}{v_{av}\tau}\right) . \qquad (3.24)$$

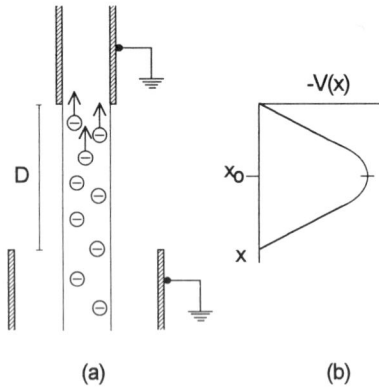

**Figure 3.4**    *(a) Charge relaxation from an emerging liquid stream back to the earthed pipe.*
*(b) The potential distribution along the stream.*

In this expression $x_0$ is the distance from the pipe outlet to the point in the liquid stream where the potential gradient changes sign (Figure 3.4b). For a narrow-necked collector whose inlet is located a distance D from the pipe outlet $x_0 \sim 0.4D$. Inclusion of this additional term predicts the maximum in $I_S$ as the conductivity increases. It also correctly predicts the velocity dependence of the maximum (Taylor, 1974) as well as the decrease in current which occurs at the higher conductivities (Figure 3.3).

The rapid decrease in $I_S$ as the liquid conductivity increases beyond about 100 pS/m offers a simple method for controlling the degree of electrification in liquids during pipeline flow.Indeed a number of commercial additives are available whose prime purpose is to increase the conductivity of the liquid so as to enhance the charge relaxation process in the outflowing liquid.

Establishing the exact concentration of additive is a matter of experimentation, and it should be remembered that the presence of the additive may affect the electrification process in the pipe as pointed out by Walmsley and Woodford (1981).

In cases where the purity of the liquid must be maintained then restriction of the flow rate remains the only viable method of static control.

### 3.1.3 Streaming Current in Transformers

A streaming current will always be generated in any system in which relatively insulating fluids are being pumped. During the last few years a number of liquid-cooled, high-voltage transformers have exploded, in some cases with catastrophic results (Crofts, 1986). The cause is believed to be the electrification of the transformer oil, the high electric field surrounding the windings in a transformer core enhancing the normal streaming current by injecting charge from the windings into the oil. The streaming current is enhanced if the oil flow becomes turbulent or its moisture content increases (CIGRW,1992),phenomena which are well-known in the flow electrification of petroleum products.

The problem can be solved by introducing additives to increase the conductivity of the transformer oil. However, because the oil is being pumped around a closed system, the choice of additive requires careful long term experimentation. As explained above, the liquid/solid interface behaves as an electrolytic half-cell and complex electrochemical process at the interface may lead to the additive plating out onto the container wall,thus rendering it ineffective.

### 3.1.4 Electrification of Polymers During Extrusion

When they are molten, polymers behave as highly viscous insulating fluids and can become electrified during extrusion in much the same way as low viscosity fluids.The degree of charging is also comparable (Taylor et al, 1974). It is unlikely that the effect will cause a hazardous situation but where extremely fine filaments of yarn are being melt-spun, sufficient charge may be generated in the polymer to cause "ballooning" effects in the yarn and strong attraction to earthed machinery. The extent of the ballooning will be determined by the degree of electrification of the polymer. As in the case of low viscosity liquids, this will depend on a number of variables,e.g. extrusion velocity and melt conductivity,which are governed by melt temperature and any additives present.

Taylor (1978) has shown that the degree of electrification may be controlled by means of an induction electrode placed near the extruder nozzle.

### 3.1.5 Streaming Current in Insulating Pipes

Electrification can also occur when liquids flow through insulating plastic pipes (Gibson 1971). With low conductivity liquids (<100 pS/m) the streaming current may be high when flow commences but as charged liquid flows out of the pipe counter charges accumulate on the inside surface of the insulating pipe. (In the case of a conducting pipe, these are discharged at the pipe wall). The additional electric fields arising from the presence of these charges reduce the tendency for further charge separation at the interface. Consequently, the streaming current decreases with time as the flow continues.

The charges on the internal surface of the pipe which are associated with the streaming current can attract adventitious, neutralising charges from the atmosphere to the outside of the pipe, giving rise to a possible hazard. Incidents have been recorded where electrical stresses across the pipe wall in such a situation increased sufficiently to cause electrical failure and subsequent mechanical rupture of the pipe. This initiated a propagating brush discharge (section 8.2.4) on the outside of the pipe which then ignited flammable liquid escaping from the pipe.

For high conductivity liquids (> 10,000 pS/m) the streaming current is constant with time and is of the same order of magnitude as that from metal pipes. Charge accumulation on the insulating pipe is believed to be limited in this case by conduction to earth through the more conductive boundary layer at the solid/liquid interface.

### 3.1.6 Other Related Electrification Mechanisms in Liquids

The formation of an interfacial double-layer can lead to electrification in a number of other industrial situations. For example, filters and strainers may result in a significant increase in the charging current because of the increase in (a) the "contact area" and (b) the turbulence. The main problem arises with the use of fine filters (< 30 μm apertures) which may readily generate streaming currents as high as 1-10 μA. The presence of even small quantities of an immiscible phase can also cause problems. For example, 0.1- 0.2 % free water in an organic liquid can increase 10- to 100- fold the streaming current from a ball valve or coarse grid.

Mixing, settling and emulsification are all processes where the shearing of an interfacial double-layer leads to the generation of static electricity.

One other example that is worthy of note is spray electrification. This occurs when a liquid stream striking a surface breaks up into fine droplets, each carrying away with it a part of the diffuse double-layer which formed at the instant of contact with the solid surface. Large electric fields and potentials may occur in the mist formed by these droplets. Indeed, the formation of a highly charged water mist during the washing of oil tankers was a crucial factor in the events that led to the violent explosions which sank one and severely damaged two very large crude oil carriers in 1969 (Van der Meer, 1971).

Naturally occurring charged mists are found at waterfalls around which the distinctive smell of ozone can often be detected.

## 3.2 CONTACT AND FRICTIONAL CHARGING OF SOLIDS

The charging that occurs when two solids come into contact has been referred to as contact charging, frictional charging, triboelectric charging and tribocharging. Usually, but by no means exclusively, contact charging is used to describe simple contacts between surfaces, i.e. contacts in which no sliding or rubbing occurs between the contacting surfaces. When any of the other terms is used it is implied that, when the contact is made, relative movement of the contacting surfaces occurs in a direction tangential to the contact interface.

At the most basic level, contact or frictional charging is a simple phenomenon. When two different but electrically neutral materials, A and B in Figure 3.5(a), come into contact, charges must transfer from one surface to the other in order to bring the contacting materials into thermodynamic equilibrium (Figure 3.5(b)). If the surfaces are separated sufficiently quickly and the two materials remain isolated from earth they will retain these excess charges (Figure 3.5(c)). Thus, by making and breaking contacts between different materials, charge separation is effected.

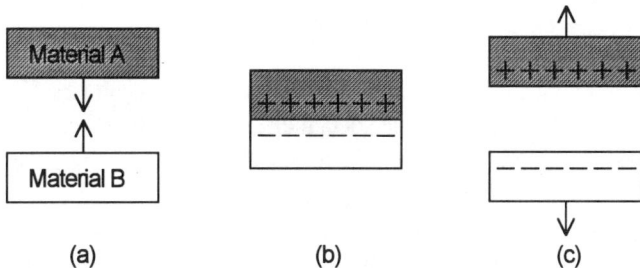

**Figure 3.5**     *The three stages of contact charging. (a) Two dissimilar materials brought into contact. (b) Charge transfers across the interface and (c) separation of the surfaces each with its excess charge.*

Contact charging occurs with all materials irrespective of whether they are good electrical conductors, e.g. metals, or good insulators, e.g. polymers. Where both materials are good conductors, the charge retained is often small because on breaking the contact charges flow rapidly to the last point of contact of the surfaces, where they recombine (Figure 3.6(a)). Even where this effect is minimised by making sphere/sphere contacts, it will be seen later that the transferred charge is reduced by back-tunnelling (in vacuum) or back-discharges (in air).

Where one or both of the materials is an insulator, charges are unable to migrate to the last point of contact because they are trapped on or just inside the insulator surface (Figure 3.6(b)). Thus charge separation is apparently much more efficient for insulators though, even here, back-tunnelling or back-discharges may reduce the overall efficiency of the process. It is this ability of insulators to trap, for significant periods of time, the excess charge transferred to their surfaces that is the main cause of static related problems in industry.

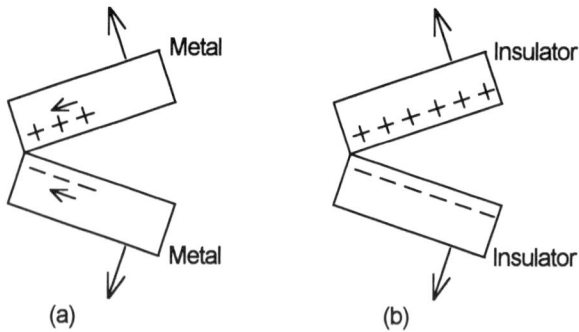

**Figure 3.6**    *(a) Charges recombining at the last point of contact as the two metal surfaces are separated. (b) Charges remain trapped on the surfaces of the two insulators and therefore cannot recombine.*

Contact charging phenomena have been studied for centuries (Gilbert, 1600; Faraday, 1843), but apart from the special case of metal/metal contacts, very little progress has been made in developing a quantitative theory for the effect. This is so even though most of the important parameters controlling the quantity of charge transferred from metals to insulators and from one insulator to another have been identified. Unfortunately, though, the many inconsistencies and contradictions in the results reported by different laboratories throughout the world militate against the development of an unambiguous theory for the separation of charge at the interface between two solids.

One major problem is the lack of consistency in the experimental technique used to investigate the phenomenon. Some measurements are carried out in air, some in vacuum. Sample preparation, sample pretreatment and surface cleaning methods all vary. Single and multiple contacts have been used as well as sliding, rolling and rubbing contacts. Pointed, spherical and flat surfaces have been investigated. Temperature, relative humidity, contact pressure, surface micro-structure and sample purity are other variables always present.

The second major problem concerns the identity of the charge species involved in the charge transfer. Harper (1967) identifies three possibilities, namely (i) electrons, (ii) ions and (iii) material transfer. The consensus is in favour of electrons, particularly where experiments are conducted under vacuum. However, some evidence for ionic effects has been presented so that ion transfer cannot be ruled out entirely when contact takes place in air. Analysis of contacting surfaces shows that sufficient material transfer can occur between contacting surfaces to explain the observed electrification. In this case, charging results from the transfer of charged radicals associated with broken bonds. Despite the evidence that material transfer occurs, few workers have taken this into account in their studies, the general belief being that it is unimportant.

The third major problem is that our understanding of the electronic structure of insulators in general and polymers in particular is still very sketchy. At best, our models are only crude approximations based on limited experimental data.

From the above discussion, it is clear that contact/frictional electrification is a complex topic requiring considerably more investigation before it is completely understood. In the following sections, therefore, the aim is to provide a general overview of the topic, concentrating on the more established features of the effect, so as to enable the reader to gain some insight into the nature of the phenomenon and of the theoretical approaches that have been adopted to explain the effect.

For those requiring a more detailed review of the subject, the text by Harper (1967) is recommended for assessing the work carried out in the period up to the mid 1960s while the more recent results and theoretical developments are analysed and evaluated in the extensive review by Lowell and Rose-Innes (1980).

### 3.2.1 Models Based on Electron Transfer

In order to describe contact charging quantitatively in terms of electron transfer, accurate knowledge is required of the electronic structure of the materials involved. The necessary information is available for most metals, for several inorganic and organic semiconductors, but only for very few insulators. Not surprisingly, models for the contact charging of insulators have evolved from a model which has proved successful in explaining the phenomenon in metals.

#### *3.2.1.1 Metal-Metal Contact*

Quantum mechanical considerations dictate that electrons in solids occupy discrete energy states. In metals, the states available for the valence electrons form a quasi-continuous band (Figure 3.7(a)) ranging from $E_C$ to the vacuum level. The vacuum level is defined as the energy of an electron removed an infinite distance away from the solid and is usually taken to be zero. The density of states per unit energy interval, N(E), increases parabolically as shown in Figure 3.7(b). The total number of states greatly exceeds the total number of valence electrons so that only those states below the Fermi energy, $E_F$ are occupied[2]. The energy interval, $\phi_W$, from $E_F$ to the vacuum level is termed the work function (or contact potential) and is an important characteristic of the metal.

---

[2]    *Strictly, this is only true at the absolute zero of temperature.*

As can be seen from Figure 3.7, $\phi_W$ represents the minimum energy that an electron in the metal must gain if it is to escape from the metal. (Those readers with a chemical background may prefer to think in terms of the electrochemical potential of the metal to which $\phi_W$ is closely related). For most metals, $\phi_W$ is in the range 4 to 5 eV, with $E_C$ some 10 to 15 eV below the vacuum level.

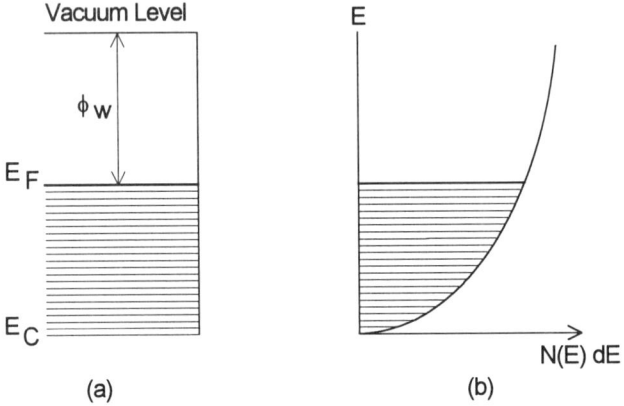

**Figure 3.7**    *(a) Simple energy band structure and (b) the density-of-states curve for a metal.*

Figure 3.8 shows a contact being made between metal A of work function $\phi_A$ and metal B with a higher work function, $\phi_B$. In (a) the two metals are far apart so that no interaction occurs between them. When the two metals are brought into contact, electrons in the higher energy states in A transfer by a tunnelling mechanism into the unoccupied lower energy states in B. Consequently, the potential of metal A becomes more positive, thereby decreasing the potential energy of all the electrons remaining in A. The potential of B becomes more negative resulting in an increase in potential energy of the electrons in B. The transfer of electrons across the junction continues until the potential difference appearing across the junction is sufficient to cause the two Fermi energies $E_F^A$ and $E_F^B$ to coincide as shown in Figure 3.8(b). This potential difference, $\Delta V$, is equal to $(\phi_B - \phi_A)/e$ volts and is known as the contact potential difference. It is equivalent to the zeta-potential at the solid-liquid interface and again a double-layer has formed (Figure 3.8(c)) albeit very thin.

On breaking the contact simultaneously at all points and assuming both metals remain isolated from earth, B retains the excess electrons gained from A and is charged negatively, while the loss of electrons causes A to be positively charged.

From detailed calculations (Harper, 1967) it can be deduced that for metal-metal contacts the charge transferred is proportional to $\Delta V$.

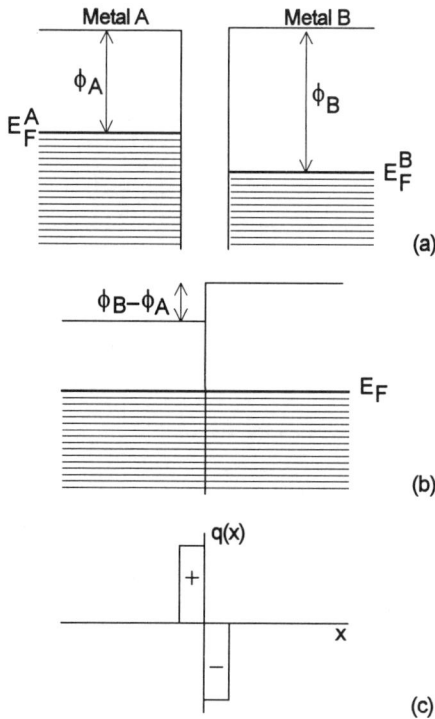

**Figure 3.8** *Contact charging of metals. (a) Two metals of different work functions brought together. (b) At thermal equilibrium the Fermi energies of the two metals are equal.(c) The charge distribution across the interface at thermal equilibrium.*

### 3.2.1.2 Metal-Insulator Contact

Charge transfer from metals to insulators can also be explained in terms of electron transfer from a solid of low work function to one of higher work function. The usual treatment considers the insulator to be a wide band-gap semiconductor with an electronic structure similar to that shown in Figure 3.9(a). Here, the energy states available to the valence electrons are split into two bands. At low temperatures, virtually all the states in the lower energy band, the valence band, are occupied by valence electrons. The upper band, the conduction band, is essentially empty but can be occupied by electrons excited from the valence band,e.g. by thermal or optical excitation. Once in the conduction band, electrons are able to contribute to the conduction process in the semiconductor.

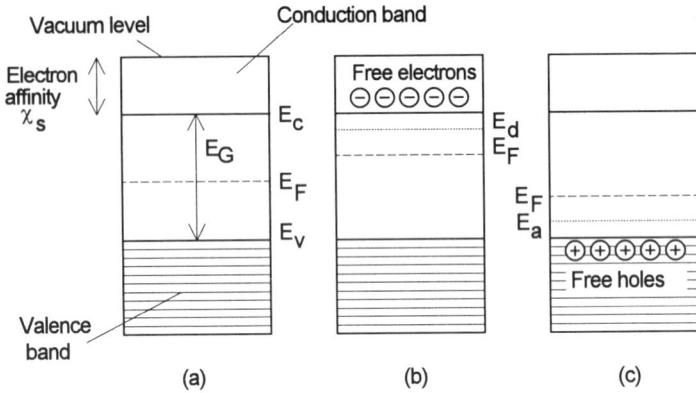

**Figure 3.9**    *Electron energy-band diagrams for (a) intrinsic, (b) n-type and (c) p-type semiconductors.*

The energy gap, $E_G$, between the valence and conduction bands is an important semiconductor parameter since it determines the intrinsic electrical conductivity, $\sigma_i$, of the material. It is readily shown (Sze,1981) that the intrinsic conductivity of a semiconductor is given by

$$\sigma_i = e(\mu_e + \mu_h)N_C N_V \exp\left(-\frac{E_G}{kT}\right) \tag{3.25}$$

where $\mu_e$ and $\mu_h$ are the electron and hole mobilities, $N_C$ and $N_V$ are the effective density of states in the conduction and valence bands, k is Boltzmann's constant and T the absolute temperature. Since $\sigma_i$ decreases rapidly with increasing $E_G$ an electrical insulator is often assumed to be a semiconductor with a wide forbidden energy gap. In silicon dioxide the band gap is about 9 eV. In polyethylene the gap is believed to be around 10 eV.

In an intrinsic (i.e. undoped) semiconductor, the Fermi energy is located in the middle of the energy gap. The work function of the semiconductor is defined in the same way as for a metal,i.e. as the energy difference between $E_F$ and the vacuum level. The electron affinity, $\chi_s$, of the semiconductor is defined as the energy difference between the bottom of the conduction band and the vacuum level.

For many insulators $\chi_S$ is so small, e.g. for polyethylene $E_C$ is believed to be close to or even slightly above the vacuum level, and $E_G$ so large that electron tunnelling to or from the valence or conduction band of the insulator is unlikely.

To explain contact charging in these materials, it was necessary therefore to invoke the presence of localised energy states in the band gap. These can arise from the presence of impurities in the semiconductor. For example, group 5 impurities such as phosphorus and arsenic introduce donor centres at an energy $E_d$ just below the conduction band in silicon as shown in Figure 3.9(b). At room temperature, electrons from these atoms are thermally excited into the conduction band where they contribute to electronic conduction. The semiconductor has been converted to n-type and it is seen that the Fermi energy is now closer to the conduction band edge than in the intrinsic semiconductor.

Group 3 impurities such as boron introduce acceptor states at an energy $E_a$ just above the valence band of silicon (Figure 3.9(c)). At room temperature, electrons from the valence band are thermally excited to these acceptor atoms leaving positively charged holes in the valence band which contribute to conduction. The semiconductor is now p-type and the Fermi energy is closer to the valence band edge. Thus, the work function of a semiconductor is determined by the type and concentration of the impurity present.

Davies (1967, 1969) interpreted insulator contact charging phenomena by assuming the insulator to be a wide band-gap, p-type semiconductor. Figure 3.10(a) shows the energy diagrams before contact is made between a metal and a semiconductor. The temperature is assumed sufficiently low that the acceptor states are empty. (This is not an unreasonable assumption since in a wide band-gap semiconductor the acceptor states are likely to be well above the valence band edge and not easily populated by thermal excitation of electrons from the valence band). On contact, electrons tunnel from the metal to the empty acceptor states. Since the density of these states is relatively low, states well away from the surface become charged. As in the case of metal/metal contacts, electron transfer continues until the Fermi energies of the metal and semiconductor are coincident. This occurs when a potential difference $\Delta V = (\phi_S - \phi_M)/e$ is built up across the interface. Applying Poisson's equation (2.23) to the space charge layer of width $\lambda$ in the semiconductor, it is readily shown (Sze, 1981) that for a constant density, $N_A$, of acceptor states

$$\lambda = \left( \frac{2\varepsilon\varepsilon_0(\phi_S - \phi_M)}{e^2 N_A} \right)^{\frac{1}{2}} \qquad (3.26)$$

and that the charge transferred per unit surface area, Q, is given by

$$Q = -eN_A\lambda = -\{2\varepsilon\varepsilon_0 N_A(\phi_S - \phi_M)\}^{\frac{1}{2}} \qquad (3.27)$$

and is proportional to the *square-root* of the contact potential difference ($\phi_S - \phi_M$).

In careful experiments conducted under vacuum, Davies (1969) found that the charge transferred depended linearly on the work function difference. This was explained by assuming that the depth of the charge layer was constant for a particular polymer but that the occupancy of the acceptor states varied with the contacting metal. Thus he was able to show that

$$Q = -\frac{2\varepsilon\varepsilon_0}{\lambda}\left(\frac{\phi_S - \phi_M}{e}\right) \qquad (3.28)$$

which is consistent with the observed linear dependence. Since for most metal/polymer combinations ($\phi_S - \phi_M$) lies between about 0.1 and 1.0 eV an explanation of the charge densities observed in Davies' experiments would require $\lambda$ to be in the range 50 - 500 nm which is large for tunnelling distances.

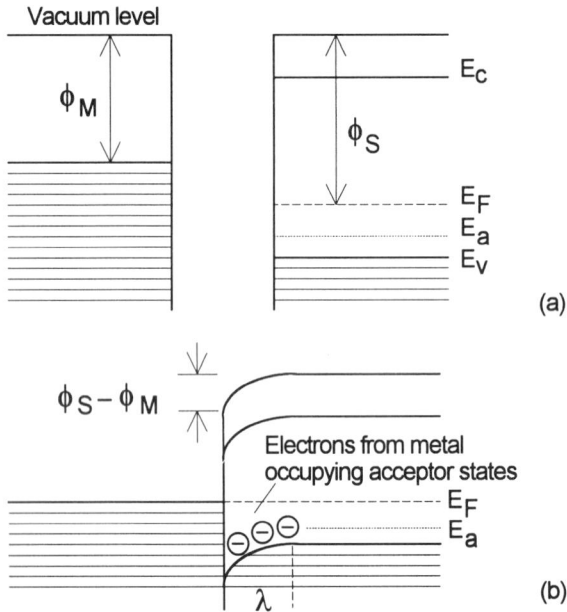

**Figure 3.10**    *Energy band diagrams of metal and semiconductor (a) before and (b) after contact.*

Chowdry and Westgate (1974a,b) argue that the assumption of a constant width for the charge layer is unreasonable, preferring instead the classic semi-conductor model leading to equation (3.27) above. Even so, they argue that a constant density of occupied acceptor sites is only realistic for a small number of insulators in which the energy of the sites is near the insulator Fermi energy. They proceed to develop a model in which the acceptor sites (or traps) are distributed uniformly in energy throughout the forbidden energy gap. The model predicts the linear dependence on metal work function observed by Davies and shows that the trap occupancy falls exponentially with distance into the insulator. Even in this model though, tunnelling distances greater than 100 nm are necessary to explain typical experimental results.

Of course, it is implicit in the above arguments that the contacting materials come into thermodynamic equilibrium (coincident Fermi energies). It is expected that equilibrium will be achieved in times of the order of the dielectric relaxation time (equation (2.59)) of the insulator. For most polymers a relaxation time is difficult to quantify since a DC conductivity is only detectable many hours after the application of very high electric fields (Taylor and Lewis, 1971) and after the transient polarisation currents have become insignificant. From such measure-ments the volume resistivity of polymers is deduced to be in excess of $10^{16}$ $\Omega$m giving relaxation times in excess of 50 hours. Most workers agree that contact charging occurs almost instantaneously (and certainly within a few seconds) although repeated contacts and longer contact times can lead to a slow increase in total charge transferred.

(a) Surface States

To overcome the relaxation time problem, many authors (Bauser et al, 1970; Bauser, 1974; Krupp, 1971) have invoked the presence of surface states on the insulator. These are electron states present in the forbidden gap in much the same way as the bulk states discussed by Chowdry and Westgate. However, as the name suggests they are localised at the insulator surface and can exchange charge rapidly on making contact to a second surface.

We must now abandon the concept of a bulk Fermi energy (that insulators of high resistivity can achieve thermal equilibrium is questionable anyway) in favour of a surface Fermi energy or neutral energy $E_n$ (Lowell and Rose-Innes, 1980). All surface states below $E_n$ are occupied by electrons while those above are empty. When contact is made to a metal, then, depending on the position of $E_n$ relative to the metal Fermi energy, $E_F$, the insulator can charge either negatively (Figure 3.11(a)) or positively (Figure 3.11(b)).

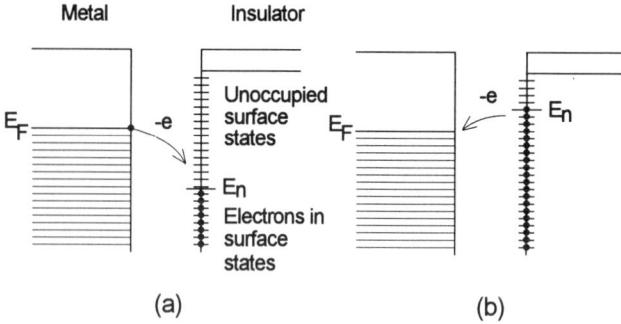

**Figure 3.11**    *Contact charging via surface states. The sign of the charge gained by the insulator depends on whether $E_n$ is (a) below or (b) above $E_F$.*

Interestingly, the dependence of the contact charge on metal work function now depends on the density and distribution in energy of the surface states. If the surface state density is low, as in Figure 3.12(a), then all states from $E_n$ to $E_F$ are filled and little change in the energy of the states occurs as a result of electrification. For a uniform distribution, $N_s$, of acceptor states per unit energy per unit area the contact charge density, Q, is given by

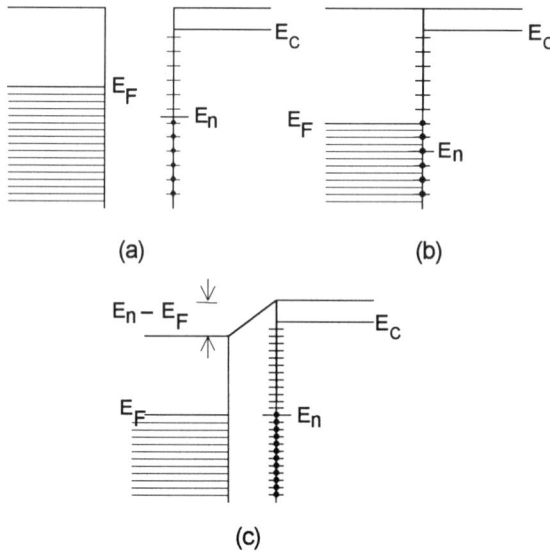

**Figure 3.12**    *(a) Metal and insulator before contact. Contact to an insulator with (b) a low and (c) a high density of surface states.*

$$Q = -eN_S(E_F - E_n) \qquad (3.29)$$

giving the often observed linear dependence on metal work function. It should be noted also that if the acceptor states are all located at the same energy, $E_A$, then the contact charge will be independent of metal work function. The states will either all be filled ($E_F > E_n$) or will all be empty ($E_F < E_n$). Results have been reported where the contact charge is independent of the metal work function (Lowell, 1976; Elsdon and Mitchell, 1976).

For other surface state distributions, $N(E)$, charge transfer will depend on the exact form of the distribution since Q will be given by

$$Q = -\int_{E_n}^{E_F} eN(E)dE. \qquad (3.30)$$

When the surface state density is high, as in Figure 3.12(b), the change in the electrostatic potential of the insulator surface is so great that a significant increase occurs in the energy of the states. As before, however, charge transfer will cease as soon as $E_n$ and $E_F$ are coincident. This will occur when the difference in potential between the metal and the polymer surface is equal to $(E_F - E_n)/e$. At equilibrium, the electric field between the metal and the polymer surface will be $Q/\varepsilon\varepsilon_0$ (equation (2.4)) where $\varepsilon$ is the effective relative permittivity at the interface. Assuming a separation d between the contacting surfaces, then

$$Q = -\varepsilon\varepsilon_0(E_F - E_n)/ed \qquad (3.31)$$

Lowell and Rose-Innes (1980) argue that, since $(E_F - E_n)$ is likely to be about 1 eV, d is of atomic dimensions, say 1 nm, and $\varepsilon$ probably about 3 in the case of a polymer, then $Q \sim 3 \times 10^{-2}$ C/m$^2$. This is a considerably greater surface charge density than has been measured experimentally and is of sufficient magnitude to cause back-tunnelling, as we will see later.

## (b) Nature of Surface States

From the foregoing discussion it is clear that surface states are strongly implicated in the contact electrification of insulators. But what are these states and where do they come from?

Surface states may be extrinsic, i.e. related to impurities in the material, or intrinsic, i.e. a property of the pure material (Lowell and Rose-Innes, 1980). In covalently bonded crystals, intrinsic states arise mainly from dangling bonds

(Shockley states) and have energies deep in the band gap. In ionic crystals the states probably arise from the perturbed orbitals of isolated ions (Tamm states) and are close to the band edge. In molecular solids where molecular interactions are weak, the states are again expected to be close to the band edges.

However, to explain experimental results requires a distribution of surface states at least 1 eV wide about the Fermi energy of metals, i.e. from about 4 to 5 eV below the vacuum level. This places the necessary states deep into the insulator band-gap. In addition, to explain positive as well as negative charging of the insulator, it is necessary to invoke the presence of donor states above $E_F$ in the insulator from which electrons can tunnel into empty metal states.

Duke and Fabish (1976) and Fabish and Duke (1978) suggest that such a distribution of states is an intrinsic property of certain side-chain polymers, e.g. polystyrene, and that surface states are simply bulk states at the surface. They argue that the normally neutral side chain moiety[3] can accept or donate electrons thus forming "molecular ions". Polarisation of the surrounding matrix, (which may occur to different extents because of the different environments in which each moiety is located) then leads to a broadening of the original state. The presence of a contacting metal can change the degree of polarisation even further (Lewis, 1990).

There is much evidence that the presence of certain chemical moieties at the surface of insulating solids can influence substantially the magnitude and polarity of the contact charge. Bauser (1974) has shown that the presence of anthraquinone on the surface of anthracene increases the negative charging of this organic crystal against most metals. Hays (1974) has shown that the contact charging of poly-ethylene by liquid mercury is small until the polymer surface is exposed to ozone. However, the most compelling evidence for the influence of chemical moieties comes from the work of Shinohara et al (1976) and Gibson et al (1979).

The Japanese workers investigated a series of p-substituted styrene polymers while the Xerox group investigated a series of substituted styrene and n-butyl methacrylate copolymers. Both groups observed a strong correlation between the polarity and magnitude of the contact charge and the Hammett substituent constant which is a measure of the electron-attracting (positive values) or electron-repelling (negative values) powers of a substituent. Table 3.2 gives values for this constant for a number of groups substituted into the *para* position in benzoic acid.

It would be expected that the more positive the Hammett constant, i.e. the more electron-attracting the moiety, then the more negative (or less positive) the poly-mer is expected to charge. Both the Japanese and Xerox groups demonstrated that

---

[3]    *A moiety is another name for a chemical group*

this was indeed the case. The presence of the amine moiety increased the positive charging tendency of a polymer while halogens and the nitro- moiety increased the tendency to negative charging.

Further evidence for the influence of chemical structure on contact charging was provided by Lowell and Brown (1988) who showed that the sign of the charge transferred from metals to polyvinyl alcohol changed from positive to negative when a fraction of the surface OH groups of the alcohol were converted to halogen (F, Cl, Br and I).

**Table 3.2**     *Some values of the Hammett constant, $\sigma_p$, for a number of p-substituted groups in benzoic acid (Finar, 1973).*

| Group | $\sigma_P$ | Group | $\sigma_P$ |
|---|---|---|---|
| $NH_2$ | - 0.66 | F | 0.06 |
| OH | - 0.37 | Cl | 0.23 |
| $OCH_3$ | - 0.27 | Br | 0.23 |
| $CH_3$ | - 0.17 | I | 0.28 |
| $C_2H_5$ | - 0.15 | $COCH_3$ | 0.50 |
| H | 0 | $NO_2$ | 0.78 |

Gibson et al (1979) further showed that charging of their polymer depended linearly on the mole percent conversion of the polymer to its substituted version. This contrasted with simple physical mixtures of identical composition where surface segregation of the components resulted in a highly non-linear dependence of charging on chemical composition.

Interestingly, the maximum charge density measured on any surface is about 1000 $\mu C/m^2$ which corresponds to about 1 electronic charge per 10 surface atoms. It seems, therefore, that only a minute fraction of the surface groups act as electron acceptors/donors.

### 3.2.1.3 Insulator - Insulator Contacts

The contacts between one insulator and another may readily be described by electron transfer in much the same way as metal-insulator contacts. Evidence for the existence of a polymer work function (equilibrium Fermi energy) has been presented by Davies (1970) thus enabling him to predict approximately the charging between specific polymers under highly controlled experimental conditions. On the other hand, Duke and Fabish (1976) provide evidence that charge transfer is controlled by surface states.

### 3.2.2 Triboelectrification

Under industrial conditions simple, non-rubbing contacts are rare; contacting surfaces come together in a way that makes a certain amount of rubbing inevitable. Experiment shows that, when an insulating surface is rubbed either by a conductor or another insulator, charge transfer may be several orders of magnitude greater than in a simple touching contact. This may be rationalised by noting that rubbing increases the intimacy of the contacting surfaces. Supporting evidence for this view has been provided by (i) Coste and Pechery (1981) who showed that charge transfer was greatest when surface roughness was small and (ii) Haenan (1976) who showed that charge transfer increased with rubbing pressure.

Generally, increasing the speed of rubbing will increase the charge transferred (Montgomery 1959), though instances have been cited where the charging goes through a maximum (Ohara, 1979) and can even change sign (Zimmer, 1970) as rubbing speed is increased. Such effects have been related to local temperature gradients appearing across the contact, resulting in the enhanced diffusion of electrons from the hotter to the cooler surface. Rubbing may also cause the surface layers of the material to be "stirred" so that charges previously on the surface are carried into the bulk (Lowell, 1976).

Despite the presence of these complicating factors, a number of workers have drawn up triboelectric series from which it is possible, within limits, to predict the polarity of the charge that is transferred from one surface to another. Two of the many series which exist (Cross, 1987) are given in Table 3.3. They are notoriously difficult to reproduce and are only useful for materials that are well-separated in the series.

The existence of the triboelectric series, coupled to the assignment of a work function to polymers, may lead one to the conclusion that no charging should occur between two surfaces made from the same material. Experiment shows this to be untrue. Significant charge transfer does occur when two notionally identical insulating materials are rubbed together. Presumably, therefore, some asymmetry must exist at the surface. Local temperature differences, material transfer and nonequilibrium occupation of surface states have all been invoked to explain the effect.

### 3.2.3 Ion Transfer

Despite the recent tendency to concentrate on electron transfer as the dominant factor in contact charging, the role of ions should not be disregarded (Harper, 1967). Charging has been found to depend on the presence of particular ionic species at the surface and in many instances protons from adsorbed moisture layers have been implicated. In the ionic model, charge transfer is seen as a

**Table 3.3**    *Triboelectric series presented by (a) Unger (1981) and (b) Hersh and Montgomery (1956). Materials at the top of the table charge positively when rubbed against those lower in the series.*

| (a) Material | (b) Material |
|---|---|
| + Asbestos | + Wool |
| Glass | Nylon |
| Mica | Viscose |
| Human hair | Cotton |
| Nylon | Silk |
| Wool | Acetate rayon |
| Fur | Lucite, Perspex |
| Lead | Polyvinyl alcohol |
| Silk | Dacron |
| Aluminium | Orlon |
| Paper | PVC |
| Cotton | Dynel |
| Steel | Velon |
| Wood | Polyethylene |
| Hard rubber | - Teflon |
| Nickel, Copper | |
| Brass, Silver | |
| Gold, Platinum | |
| Sulphur | |
| Acetate rayon | |
| Polyester | |
| Celluloid | |
| Orlon | |
| Saran | |
| Polyurethane | |
| Polyethylene | |
| Polypropylene | |
| PVC | |
| Silicon | |
| - Teflon | |

redistribution of ions between the contacting surfaces. However, since we have little information on the ionic population of surfaces before contact is made, it is not possible to predict the magnitude of the transferred charge.

### 3.2.4 Role of Back-Discharges and Back-Tunnelling in Tribocharging

Allusion has been made already to the difficulty of obtaining reproducible experimental results when attempting to measure triboelectrically generated charges even under rigorously controlled experimental conditions. The inability to predict the magnitude of the transferred charge is also troublesome. It is reasonable to question, therefore, whether the results obtained in such experiments are simply artefacts of the experiment. An indication that this may be true in some cases comes from a comparison of tribocharging experiments carried out in air, where the magnitude of the tribocharge is in the range 1-100 $\mu C/m^2$, with those in vacuum where the transferred charge increases to about 10 -1000 $\mu C/m^2$.

In the air experiment when the contacting surfaces separate, the electric field in the gap may be sufficiently great to cause electrical breakdown of the air in the gap. The discharge will deposit neutralising ions onto the respective surfaces as shown in Figure 3.13 thus leading to a reduction in the total charge transferred. The occurrence of such discharges may be inferred from the appearance of "crows-foot" patterns (Lichtenberg patterns) that develop as dust accumulates on plastic surfaces. They are particularly visible on the plastic surfaces of consumer white goods which have been in storage for some time.

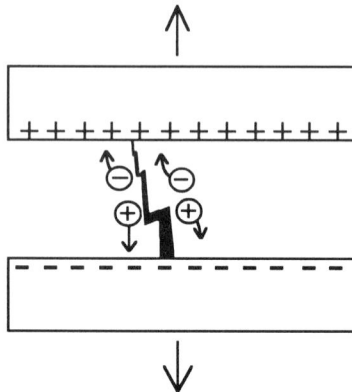

**Figure 3.13**    *A back-discharge in air between separating surfaces sprays neutralising ions onto charged surfaces.*

Gas discharges are statistical in nature, and so it should not come as a surprise, therefore, that results obtained in tribocharging measurements may have a considerable statistical spread.

When experiments are conducted under vacuum, the limitation imposed by the relatively low electrical breakdown strength of air is removed. Consequently, higher fields may be tolerated in the gap between the separating surfaces. Now a second limiting mechanism comes into play, i.e. back-tunnelling. Consider the situation in Figure 3.14 in which N electrons per unit area have tunnelled from the metal to a set of unoccupied surface states on the insulator, located $\Delta E$ below the metal Fermi energy. As the surfaces separate, the transferred charge sets up an electric field of magnitude $Ne/\varepsilon_0$ in the gap, which in turn gives rise to a potential difference V between the metal and the polymer surface where V is given by

$$V = Nex/\varepsilon_0$$

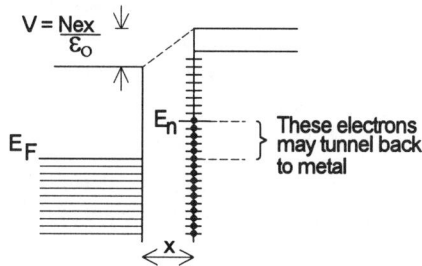

**Figure 3.14**   *Separating charged surfaces creates a potential difference which changes the energies of the electrons in the insulator relative to those in the metal. Electrons in states above $E_F$ can tunnel back to the metal if $x < 5$ nm.*

This potential difference raises the energy of the electrons in the surface states by an amount $Ne^2x/\varepsilon_0$. When the separation of the surfaces has increased to some critical distance $x = x_0$ the energy of the electrons trapped in the surface states of the polymer will be greater than the Fermi energy of the metal. Consequently, there will be a tendency for these electrons to tunnel back into the empty, lower energy states in the metal.

It is generally accepted that tunnelling probabilities are almost negligible for distances greater than about 5 nm and $\Delta E$ normally will be of the order of 1eV. We can now estimate the maximum charge that may be transferred since

$$Ne = \varepsilon_0 \Delta E/ex_0 \approx 1.8 \times 10^{-3} \text{ C/m}^2$$

which is somewhat greater than the surface charge-density values observed experimentally in vacuum. This supports the idea that back-tunnelling can indeed occur in measurements carried out under vacuum.

In summary, it is likely that,for clean surfaces coming into contact in vacuum, the charge transfer is controlled initially by the density of surface states and the effective work function of the insulator. However, as the surfaces separate, back-tunnelling partially neutralises the charges on the surfaces. When contacts are made in air, the initial charging mechanism may involve either electrons or ions or both. In all cases, though, the final charge may be reduced significantly by a discharge in the air gap which opens up between the separating surfaces. All these effects are complicated by a host of possible phenomena including differences in contact area, distortion of the sample surface, local temperature differences etc. In our present state of knowledge, predicting how much charge is generated by triboelectrification is extremely imprecise.

### 3.2.5 Tribocharging Limits in Industrial Processes

Although current theories cannot predict the magnitude of the charge that will be generated in industrial situations we can, nevertheless, set upper limits for various situations. This is so because,above some critical surface charge density, the electrical breakdown field of the ambient air (2.7 MV/m) will be exceeded. A corona, a brush or even a spark discharge may then occur which will dissipate the excess surface charge. Two important limits are given below.

(a) Insulating Sheet

In Chapter 2 it was seen that an electric field of magnitude $Q/2\varepsilon_0$ was directed away from either side of an isolated sheet charged uniformly to a surface density $Q$ per unit area. The maximum charge density, $Q_{MAX}$, that can be tolerated before this field exceeds the breakdown strength, $E_{BD}$, of the ambient medium is given by

$$Q_{MAX} = 2\varepsilon_0 E_{BD}. \tag{3.32}$$

Where one surface of an initially uncharged sheet is tribocharged, the limiting charge density will be half this value because,as the contacting surfaces separate, all the flux lines, i.e. the electric field, will be bounded by the two separating surfaces as shown in Figure 3.15. Thus

$$Q_{MAX} = \varepsilon_0 E_{BD} \tag{3.33}$$

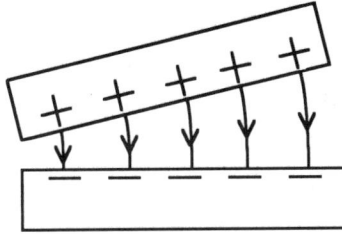

**Figure 3.15**    *When tribocharged surfaces remain close together the electric field is confined within the space between the surfaces.*

which for air yields

$$Q_{MAX} = \pm 24 \ \mu C/m^2.$$

When sheet products are being handled, e.g. passing through a system of rollers, it is possible for the charge density on the individual surfaces, $Q_1$ and $Q_2$ respectively, to exceed this limit so long as the net charge density $[Q_1 + Q_2]$ remains less than $Q_{MAX}$. Such a situation could arise if a slipping rubber roller generates a large tribocharge on the upper surface of the sheet in Figure 3.16. While the sheet remains in contact with the lower metal roller, this high charge density can be sustained because of the greater breakdown strength of the sheet compared with air. Once the sheet comes away from the roller, the field in the air gap exceeds the breakdown field of air and a discharge is initiated. Negative charges from the discharge are drawn to the lower surface of the sheet thus reducing the net charge on the sheet below the air limit once again.

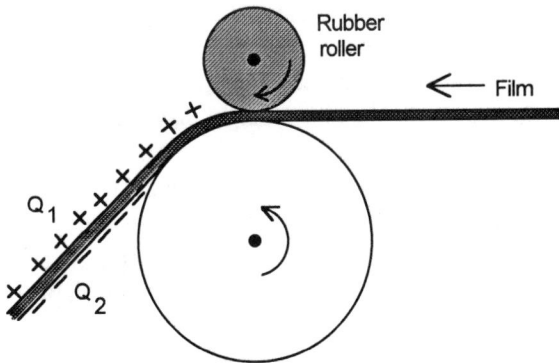

**Figure 3.16**    *Mechanism for generating large surface charges but low net charge on a film.*

We must remember,though, that ,while the net charge on such a film may be low, high charge densities of opposite polarity have been created on opposite surfaces of the sheet. This is one of the conditions required for the initiation of dangerous propagating brush discharges (section 8.2.4).

### (b) Spherical Particles

For small isolated particles, air breakdown again limits the maximum surface charge density to $Q_{MAX}$ ($= \varepsilon_0 E$) which corresponds to a maximum charge, $q_M$, on a particle of radius r given by

$$q_M = 4\pi r^2 \varepsilon_0 E_{BD}.$$

However, when estimating the limiting charge in this case it must be remembered that the limiting field for corona onset is also a function of the radius of curvature of the stressed surface (Cobine, 1958). For a spherical particle of radius r, the critical field is given by

$$E_C = \frac{0.18}{\sqrt{r}} \qquad\qquad (3.34)$$

where $E_c$ is in MV/m and r in metres.The limiting surface charge density in $\mu C/m^2$ is then given by

$$Q_{MAX} = \frac{1.6}{\sqrt{r}}. \qquad\qquad (3.35)$$

For a particle of radius 100 $\mu$m and density 1000 kg/m³ this corresponds to a specific charge of 4800 $\mu$C/kg.

This limit for an individual particle is rarely applicable in powder handling processes because of bulking effects. When a powder is concentrated, corona is initiated when the cumulative field of all the particles reaches the air breakdown limit. For simple cases the limiting charge may be determined using Gauss' theorem (equation (2.9)). As an example, assume that the powder discussed above is being transported pneumatically through a cylindrical metal pipe of diameter D and is suspended within the pipe at a uniform concentration of 0.1 kg/m³. From section 2.3.4.3 the maximum volume charge density $Q_M$ which can be tolerated in the pipe before corona is initiated at the pipe wall is given by

$$Q_M = \frac{4\varepsilon_0 E_{BD}}{D} \qquad\qquad (3.36)$$

which, for a 1 m diameter pipe,gives a maximum specific charge on the powder of 956 µC/kg. When the same powder is packed into a cylindrical metal bin of the same diameter in which the packing density is, say, 500 kg/m³ the maximum specific charge reduces even further to 0.19 µC/kg.

In both cases, the excess charge is transferred by air discharges from the particles to the wall of the pipe or bin. If, as is often the practice, the bin is lined with an insulating bag, the excess charges will accumulate on the inside surface of that bag creating a possible electrostatic hazard when it is ultimately removed from the bin.

### 3.2.6  Tribocharging in Industry

#### (a) Applications

Tribocharging has a notorious reputation for fickleness. Nevertheless, a number of industrial processes have evolved which rely on this form of charging. Tribo-electric spray guns are increasingly being used in preference to corona guns for dry powder coating in order to minimise the disruption of the deposited layer caused by back-ionisation effects (section 3.4.2.5). These result from the excess charging of the coated surface by air ions created in the corona source. From the review of tribocharging above, it is clear that certain powder formulations will charge better and more reproducibly than others. Unfortunately, the process may be very sensitive to the presence of impurities and to changes in the relative humidity of the ambient.

The most consistently reliable application of tribocharging is in electrophoto-graphy (see Chapter 1) where the latent image on the surface of the photoconducting drum is developed by pouring charged toner particles over the drum surface. To ensure that the toner is attracted to, and faithfully reproduces, the latent image it is essential that the charge on the toner particles is of opposite polarity to that forming the image. This is achieved triboelectrically by mixing the toner with carrier beads. As with powder coating technology, the precise formulation of toner/carrier systems is important in determining the success of the process. The warm, dry atmosphere within an operational photocopier is also important; by maintaining a controlled environment in which to carry out the tribocharging, consistent performance is achieved.

#### (b) Hazards and Problems

Tribocharging is possibly the greatest source of unwanted static electricity in industry and is likely, therefore, to be at the root of most electrostatic problems. Since it is not possible to predict theoretically the magnitude or indeed in many cases even the polarity of the charge, experimental measurements must play a leading role in analysing a particular problem, as was the case with the streaming current in the petrochemical industry.

In powder handling operations it is generally found that the greatest charge is associated with the finest powders. It seems also that, for organic powders at least, the specific charge on the powder depends more on the operation being carried out rather than the formulation of the powder. Examples of typical operations and the range of specific charge that can be expected are given in Table 3.4.

Film winding operations are invariably prone to tribocharging since the film surface is continually making and breaking contact with metal and/or rubber rollers and often with itself. Sticking rollers which cause the film to slip exacerbate the problem. Increased winding speed and contact pressure can also increase tribocharging. In most cases the surface charge density will not be constant. It may even change sign! This probably indicates the presence of back-ionisation when the film separates from the roller. The discharge may vary in intensity from a low level corona to a severe spark depending on the rate of charge generation, film speed and the distance between adjacent rollers (Nickell and Taylor, 1991). Nevertheless, it should be possible to overcome most static problems associated with film winding operations by the correct use of static eliminators (Secker, 1979).

**Table 3.4**    *The specific charge that can be generated during various powder handling operations.*

| Operation | Specific Charge ($\mu$C/kg) |
|---|---|
| Sieving | $10^{-5}$ - $10^{-3}$ |
| Pouring | $10^{-3}$ - $10^{-1}$ |
| Scroll feed transfer | 0.01 - 1 |
| Grinding | 0.1 - 1 |
| Micronising | 0.1 - 100 |
| Pneumatic transfer | 1 - 100 |

## 3.3 CHARGING BY INDUCTION

It was shown in Chapter 2 that real charges induce image charges in adjacent conductors. In later chapters we will see how this effect is utilised in a number of devices for measuring static charge, for example, the Faraday pail and field mill.

The phenomenon is also particularly useful for the controlled charging of low resistivity materials, i.e. materials with volume resistivity $\leq 10^8\ \Omega\text{m}$.

The sequence of events leading to the generation of charge by induction is shown in Figure 3.17. In (a) a low resistivity, electrically neutral particle arrives in the space between two electrodes across which a voltage is applied. The electric field in this space causes charges within the particle to redistribute in order to maintain a constant potential throughout its volume. Positive charges are induced on the side facing the negative electrode, negative charges on the side facing the positive electrode. The particle is said to have polarised. If, now, the particle momentarily contacts the earthed electrode, negative charges flow out of the particle into the electrode and neutralise an equal number of positive charges in the electrode. The particle is left with an excess positive charge (Figure 3.17(b)) and when it breaks away from the electrode this excess charge is retained (Figure 3.17(c)). It should be remembered that had the particle touched the negative electrode it would have acquired a negative charge.

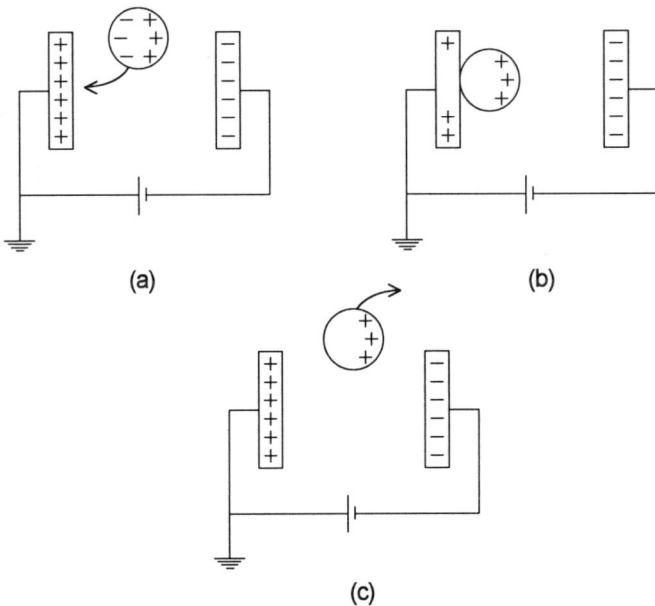

(a)

(b)

(c)

**Figure 3.17**   *The three stages in induction charging. (a) Polarisation in an electric field, (b) charge transfer and (c) charged particle moves away from charging electrode.*

Once in contact with the electrode, the time taken for a particle to acquire its equilibrium charge is determined by its dielectric relaxation time, $\tau$ (equation (2.59)) and depends, therefore, on its resistivity. A resistivity of $\leq 10^8$ $\Omega$m will ensure that charging takes place in $\leq 1$ ms, which is sufficiently rapid for most industrial applications.

The various stages of induction charging may be observed by watching closely the behaviour of airborne dust particles in the space between a pair of electrodes across which a high voltage is applied. When a particle has charged at one of the electrodes, as described above, the electric field will cause it to break away from the surface and migrate to the counter electrode. There, the particle loses its excess positive charge, gains an excess negative charge and migrates back to the positive electrode, where the whole cycle repeats. Similar behaviour has been observed between electrically stressed electrodes immersed in insulating liquids.

Any low resistivity particle contacting an electrode, in the presence of an electric field, will become charged by induction and the phenomenon has found applications in a number of industrial processes and devices. Under certain circumstances it can also lead to the inadvertent charging of personnel.

### 3.3.1 Ink-Jet Printers

In ink-jet printers the "particle" is a small ink droplet ejected from a small orifice in one of the electrodes (Figure 3.18) which also serves as a reservoir. In one type of printer, the droplet size and the frequency of ejection from the orifice are controlled by pressure waves from a small piezoelectric transducer. As it emerges from the orifice a charge is induced on the droplet which is opposite in

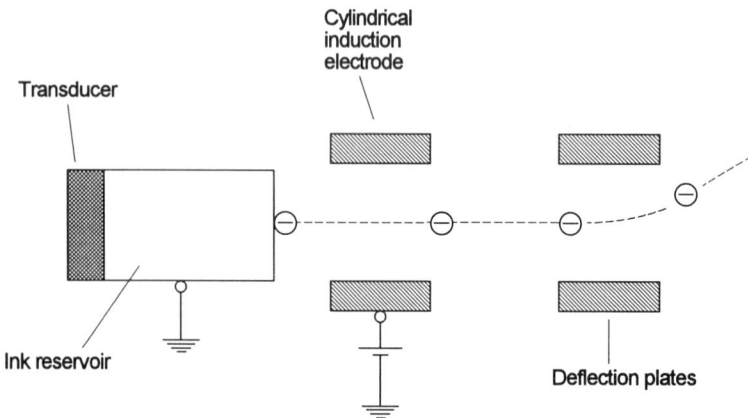

**Figure 3.18**    *Basic elements of an ink-jet printer.*

sign to that on the induction electrode located downstream. The stream of charged droplets accelerated through this annular electrode may then be deflected to a predetermined position by applying voltages across two pairs of deflection plates in much the same way as the electron beam is deflected in a cathode-ray tube. When a blank pixel is required the droplets are directed into a "gutter".

The charge, $Q$, acquired by a particle or droplet in contact with an electrode is determined by the usual capacitance equation. To a first approximation the particle may be treated as an isolated sphere of radius R with a capacitance equal to $4\pi\varepsilon_0 R$ so that

$$Q = 4\pi\varepsilon_0 RV \qquad (3.37)$$

where $V$ is the voltage between the electrodes.

### 3.3.2 Atomisation

Particles charged by induction are subjected to the same corona limit as those charged triboelectrically. For liquid droplets, though, a second and much more important limit may well be reached before corona sets in. Consider a droplet of radius R connected (via a fine capillary) to a high voltage source as shown in Figure 3.19. The electrostatic energy, U, stored in the droplet is readily calculated since

$$U = \frac{1}{2}CV^2 = \frac{1}{2} \times 4\pi\varepsilon_0 R \times V^2.$$

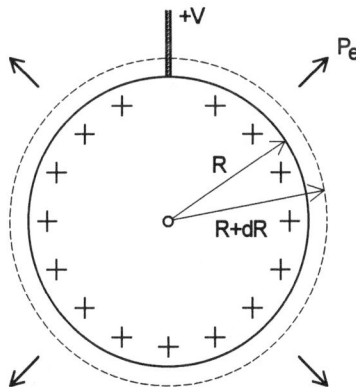

**Figure 3.19**    *Mutual repulsion of electrostatic charges causes droplets to expand. This is equivalent to an electrostatic pressure $P_e$.*

The mutual repulsion of the charges on the droplet will cause it to expand. In other words, applying a voltage to the droplet has created an electrostatic pressure, $P_e$, within it. By equating the work done by this pressure in expanding the droplet from a radius R to R+dR to the increase in electrostatic energy arising from the change in droplet capacitance we may calculate the magnitude of the pressure. Thus,

$$P_e \times 4\pi R^2 dR = \frac{1}{2} \times 4\pi\varepsilon_0 dR \times V^2$$

or

$$P_e = \frac{\varepsilon_0 V^2}{2R^2}. \tag{3.38}$$

Now assume that this pressure is attempting to split the droplet along a diametrical plane against the surface tension, $\gamma$, which is trying to hold the droplet together. If the droplet is to remain stable, then the electrostatic force must remain less than the surface tension force, i.e.

$$P_e \times \pi R^2 \leq 2\pi R\gamma.$$

Substituting for P from equation (3.38) we arrive at the Rayleigh limit for a stable droplet, which is expressed as

$$V^2 \leq \frac{4\gamma R}{\varepsilon_0}. \tag{3.39}$$

When V increases above the limit set by equation (3.39) the electrostatic forces exceed the surface tension, instability sets in and the droplet atomises into a fine mist, which has the effect of reducing the total energy of the system. Following Hines (1966) the radius r of a mist droplet after atomisation is given by

$$r^3 = \frac{9\gamma\varepsilon_0}{Q^2} \tag{3.40}$$

where $Q$ is the volume charge density in the droplet. In a series of tests with mixtures of alcohol and water he was able to show that experiment was in close agreement with theory; the droplet sizes ranging from 25μm < r < 83 μm for charge densities in the range 2.1 C/m³ < $Q$ < 7.9 C/m³.

Electrostatic atomisation of liquids is used in wet painting and in agriculture. In such applications, the atomisation takes place at a nozzle raised to a high voltage.

As soon as they are formed, the charged droplets drift in the electric field between the nozzle and the earthed surface being sprayed. Electrostatic spraying comes into its own when irregularly shaped objects are to be painted, since charged droplets will follow the electric fields into re-entrant cavities and also to the under and rear surfaces of the workpiece. This is particularly beneficial in crop spraying since the underside of leaves, where the pests and fungal spores are to be found, are coated almost as efficiently as the upper surface. "Mechanical" sprays reach only the upper surface of the leaves.

Electrostatic atomisation is used also in nebulisers for more efficient delivery of medication into the lungs.

### 3.3.3  Induction Charging of Personnel

In the above examples, the electric field for charging the particles and droplets was deliberately created by connecting an electrode to a high voltage supply. The source of the inducing electric field is immaterial though and can just as easily be created by static charges trapped on insulating surfaces.

A typical industrial example is illustrated in Figure 3.20 where a negatively charged plastic film is  being collected on a wind-up reel. The high charge density in such a reel will set up an electric field in the space between it and the nearby earthed surfaces. An operator wearing insulating footwear walking into this space will polarise in exactly the same way as the particle in Figure 3.17 because the human body is highly conductive. Positive charges will be attracted to one side of

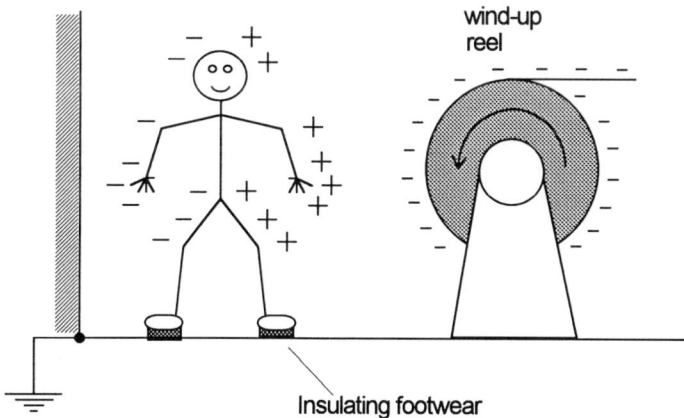

**Figure 3.20**  *Polarisation of personnel near a charged wind-up reel. On touching an earthed surface in this location the person will charge positively by induction.*

the operator's body while negative charges will be repelled to the remoter parts. Should the operator now approach and touch an earthed surface while in the presence of this charged reel, negative charges will flow from his/her body to earth leaving him/her positively charged. Occasionally, operators will experience severe electrical shocks during this stage; sufficient in some cases to cause loss of conciousness.

When the operator walks away from the charging location to perform other duties, the excess charge will be carried on the body and will discharge when another earthed surface is touched. This again may be a painful experience!

When the operator returns to the vicinity of the charged reel, the whole process will repeat. In the presence of highly charged reels, it is not unknown for the body potential of operators to increase sufficiently for the electrical strength of some types of footwear to be exceeded, resulting in a tingling feeling in the operator's feet as he discharges to earth through his shoes!

Problems such as those described above are best avoided by using neutralising systems to prevent static build-up at the wind-up reel and ensuring that operators wear conductive footwear. It should be noted though that wearing conductive footwear without a neutralising system could lead to operators receiving severe shocks directly from the reel.

## 3.4 CORONA CHARGING

In high voltage engineering, the occurrence of a corona discharge is often a portent of serious electrical failure in the system. By contrast, in a wide range of industries corona discharges are deliberately initiated and utilised in a number of applications, particularly for the precipitation of dusts and smoke and for dry powder coating. Such applications are only possible because corona discharges are highly controllable sources of both positive and negative electrostatic charges.

### 3.4.1 The Corona Discharge

A corona discharge is a self-sustaining, partial breakdown of a gas subjected to a highly divergent electric field such as that arising near the point in the point-plane electrode geometry shown in Fig. 3.21(a). In such an arrangement, the electric field, $E_p$, at the corona point is considerably higher than elsewhere in the gap. To a reasonable approximation $E_p$ is independent of the gap between the electrodes and given by

$$E_P = \frac{V}{r} \tag{3.41}$$

where V is the potential difference between the point and plane and r is the radius of the point. As V increases, the electric field in the vicinity of the point becomes sufficiently great to trigger an electron avalanche. This is the so-called Townsend α-process, in which the electron population grows exponentially as a result of ionising collisions with neutral gas molecules. Although a fraction of the secondary electrons attach to neutral molecules forming negative ions, the rapid growth in the free electron population quickly leads to breakdown (section 8.1).

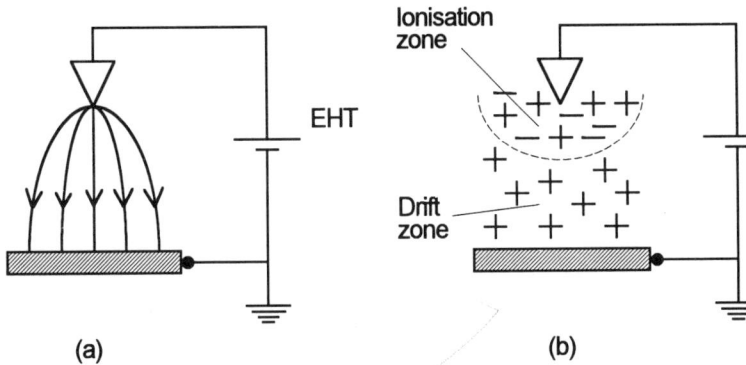

**Figure 3.21**     *(a) Non-uniform field in a point-plane electrode geometry. (b) The two main zones in a corona discharge.*

In a corona discharge the breakdown is confined to the high field region near the point, resulting in a highly ionised region there (Figure 3.21(b)). This ionisation zone is usually characterised by a faint glow, the colour and appearance of which depend on the gas and on the magnitude and polarity of the voltage applied to the point. The occurrence of a glow indicates that there must also be short-lived excited species present in the ionisation zone in addition to the high concentrations of positive ions, negative ions and free electrons. When they relax to the ground state, the excited species emit photons whose wavelengths are characteristic of the gas in which the corona is occurring.

Complete electrical failure of the gap is prevented by the existence of the drift zone which effectively provides a stabilising resistance in series with the ionisation zone. No ionisation takes place in the drift zone because the electric field here is below the critical value required to accelerate free electrons to ionisation energies between successive collisions with gas molecules.

When positive voltages are applied to the point, negative ions and electrons are accelerated towards the point while positive ions drift out of the ionisation zone and flow through the drift zone towards the planar electrode. The unipolar current flow in the drift zone should be contrasted with the bipolar current flow in the ionisation zone. When negative voltages are applied to the point, negative ions now drift out of the ionisation zone and flow through the drift zone to the planar electrode.

It is the unipolar ion flow in the drift zone that is utilised in corona charging since objects placed in the drift zone will intercept the ion stream and become charged. By applying a voltage of appropriate polarity to the point electrode both positive and negative charging may be effected.

*3.4.1.1 Corona Threshold Voltage*

Before a corona is initiated in air the threshold field given by equation (3.34) must be exceeded at the point. Substituting into equation (3.41) then gives the critical voltage, $V_C$ in volts, for corona onset as

$$V_C = 1.8 \times 10^5 \sqrt{r} \qquad (3.42)$$

for a point with radius r (in metres). Values for corona onset voltages calculated from equation (3.42) for points of different radii are given in Table 3.5.

**Table 3.5**    *Values calculated from equation (3.42) for the corona onset voltage for a series of points of different radii.*

| Radius of point (mm) | 1 | 0.5 | 0.1 | 0.05 |
|---|---|---|---|---|
| $V_C$ (kV) | 5.69 | 4.02 | 1.8 | 1.27 |

In a number of applications, e.g. electrostatic precipitation and electrophotography, the corona source is a wire. If the corona wire is held coaxially within an earthed cylinder the field, $E_W$, at the wire is given by

$$E_W = \frac{V}{r \ln (R/r)} \qquad (3.43)$$

where r is the radius of the wire, R the radius of the outer earthed cylinder and V the applied voltage. The above equation is also a good approximation for a wire-plane arrangement. The threshold field for corona initiation at a wire in air has been given by Peek (1929) as

$$E_C = (f\delta)\left(A + \frac{B}{\sqrt{\delta r}}\right) \qquad (3.44)$$

where $A = 3.1 \times 10^6$ V/m, $B = 9.55 \times 10^4$ V/m$^{1/2}$, r is in metres, f is a roughness factor ($<1$) and $\delta$ is the relative air density (equal to 1 at 25°C and 1 atmosphere). Combining equations (3.43) and (3.44) yields the corona onset voltage $V_C$. Thus

$$V_C = (f\delta r)\left\{A + \frac{B}{\sqrt{\delta r}}\right\}\ln\frac{R}{r}. \qquad (3.45)$$

For a smooth wire operating in a normal atmosphere both f and $\delta$ are close to unity, so that

$$V_C \approx \{Ar + B\sqrt{r}\}\ln\frac{R}{r}. \qquad (3.46)$$

Table 3.6 gives the calculated corona thresholds in air for smooth wires of various radii and for R=10 mm and 50 mm. For both the point-plane and wire-plane arrangements, corona threshold voltages calculated from equations (3.42) and (3.46) underestimate the values obtained experimentally for radii smaller than 0.05 mm. For fine wires positioned 12.7 mm above an earthed plane in air Vyverberg (1965) has suggested that the empirical relation

$$V_C = 2.85 + 34.6r \qquad (3.47)$$

should be used to calculate V in kV if r is in mm.

It should be noted that,when a corona is initiated, the average field in the gap is significantly smaller than the breakdown field of the gas.

**Table 3.6**    *Corona thresholds for two different air gaps, calculated from equation (3.46) for smooth wires of different radii in normal laboratory air.*

| Wire radius (mm) | Corona threshold voltage (kV) | |
|---|---|---|
| | (R = 10 mm) | (R = 50 mm) |
| 1.0 | 14.1 | 23.9 |
| 0.5 | 11.0 | 17.0 |
| 0.1 | 5.8 | 7.9 |
| 0.05 | 4.4 | 5.7 |

*3.4.1.2 Current-Voltage Characteristic of a Corona*
    The unipolar ion current I per unit length flowing from a wire of radius a held coaxially within an earthed cylinder of radius R is given by

$$I = 2\pi r Q(r)\mu E(r) \qquad (3.48)$$

where $Q(r)$ and $E(r)$ are respectively the volume charge density and electric field at a radial distance r from the wire axis and $\mu$ is the mobility of the ions contributing to the current. Poisson's equation (equation 2.23) in cylindrical co-ordinates

$$\frac{1}{r}\frac{d}{dr}(rE(r)) = \frac{Q(r)}{\varepsilon_0} \qquad (3.49)$$

provides a second relationship between $Q(r)$ and $E(r)$ for this geometry. Substituting from equation (3.48) for $Q(r)$, assuming $E(a)=E_0$ and integrating yields

$$E(r) = \left[\frac{I}{2\pi\varepsilon_0\mu} + \left(\frac{a}{r}\right)^2\left(E_0^2 - \frac{I}{2\pi\varepsilon_0\mu}\right)\right]^{\frac{1}{2}}. \qquad (3.50)$$

As can readily be deduced when $r \gg a$, i.e. a long way from the wire, equation (3.50) simplifies to

$$E(r) = \left(\frac{I}{2\pi\varepsilon_0\mu}\right)^{\frac{1}{2}} \qquad (3.51)$$

showing that the field becomes constant and independent of position. Townsend (1914) has shown that, for small currents, equation (3.50) may be further integrated yielding

$$I = KV(V - V_C) \qquad (3.52)$$

where K is a constant given by

$$K = \frac{8\pi\varepsilon_0\mu}{R^2\ln(R/a)}. \qquad (3.53)$$

In these equations V is the applied voltage and $V_C$ the corona onset voltage. When large currents flow, the approximations made to obtain equation (3.52) are no longer valid and a more complex relation holds (Cobine 1958).

A similar derivation for the current-voltage characteristic of the point-plane corona is not possible; the mathematics rapidly becomes intractable. Nevertheless, it has been shown by experiment (Warburg, 1899; Lamo and Gallo, 1974) that a similar expression to equation (3.52) still holds but with K now dependent on the gap and the radius of the point.

Although the arguments above suggest that the current-voltage characteristics of positive and negative coronas are identical, in practice this is not the case. Experimental data for point-plane geometry (Cobine, 1958) show a slightly lower onset voltage for positive corona in air but the current-voltage curves cross at higher voltages. For wires of diameter greater than 0.075 mm positive corona again appears before negative corona but the reverse is true for smaller diameter wires.

### 3.4.1.3 Current Density in a Corona

For the case of a wire corona, cylindrical symmetry ensures a uniform ion current density arriving at the outer cylinder. In the point-plane corona this is not the case. The current density at the planar electrode is greatest immediately under the point and decreases elsewhere. Warburg (1928) gives the current density at the plane as

$$j = j_0 \cos^m(\theta) \tag{3.54}$$

where $j_0$ is the current density immediately below the point, m is 4.65 for positive corona and 4.82 for negative corona and $\theta$ is the angle between the axis and a line drawn from the point to a position of interest on the plane as shown in Figure 3.22.

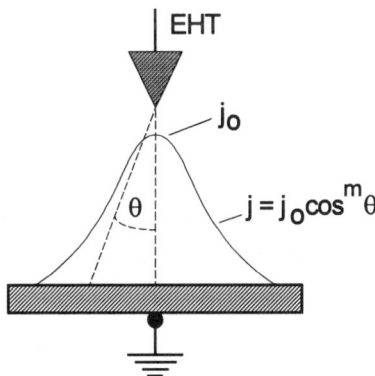

**Figure 3.22**    *Distribution in the current density of a point-plane corona.*

### 3.4.2 Corona Charging of Airborne Particles

Airborne particles entering the drift zone of a corona discharge will intercept the unipolar ion flow and, as a consequence, will become charged. To calculate the magnitude of the charge acquired by the particle it is necessary first of all to determine the distortion of the electric field caused by induced and real charges on the particle. In the absence of the particle, equation (3.51) shows that the field in the drift zone may be assumed constant. Initially we shall ignore the charging effects of the ion current. When a spherical particle is placed in the drift zone, the previously uniform electric field distorts, the flux lines bending towards the particle ($\varepsilon > 1$) as shown in Figure 3.23. This symmetrical field pattern, produced by polarisation within the particle, is distorted when ions are intercepted and trapped on the surface of the particle. Initially, charges build up on the surface facing the ion source but if the particle is reasonably conductive, the intercepted charge is distributed rapidly over the whole surface. For insulating particles this may not be possible. Nevertheless, because of the electrical torque acting on the non-uniform surface charge, the particle will begin to spin randomly, thus ensuring uniform charging of the surface even in this case.

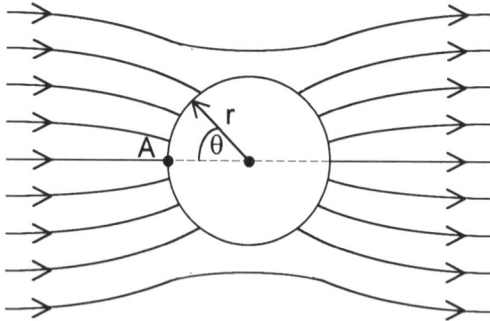

**Figure 3.23**    *Polarisation of a spherical particle causes a distortion in the previously uniform electric field.*

In both cases, charging will continue until the electric field of the charges trapped on the particle is sufficient to prevent any more ions from reaching the particle (Figure 3.24).

### *3.4.2.1 Limiting Charge*

To calculate the magnitude of the limiting charge, it is necessary first to determine the electric field at the particle surface just before charging commences. This is achieved by solving the Laplace equation (equation (2.24)) which in the spheric-

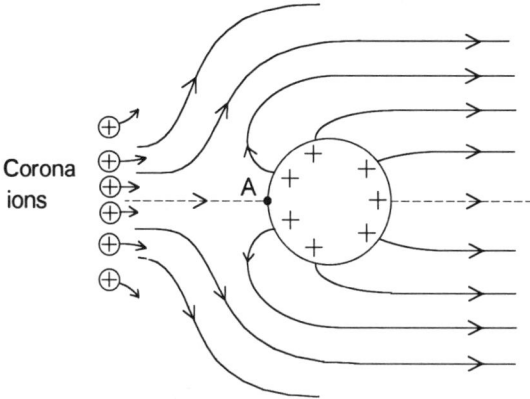

**Figure 3.24**    *Field pattern around a fully charged particle. The field at A is zero so that the ion flow is deflected around the particle.*

al coordinate system required here (Figure 3.23) is written as (Smythe, 1968)

$$r^{-2}\frac{\partial}{\partial r}\left(r^2\frac{\partial V}{\partial r}\right) + \frac{1}{r^2\sin\theta}\frac{\partial}{\partial\theta}\left(\sin\theta\frac{\partial V}{\partial\theta}\right) + \frac{1}{r^2\sin^2\theta}\frac{\partial^2 V}{\partial\phi^2} = 0 \ . \qquad (3.55)$$

This equation has solutions of the general form

$$V_o = \left(\frac{A}{r^2} + Br\right)\cos\theta \qquad (3.56a)$$

and

$$V_i = \left(\frac{C}{r^2} + Dr\right)\cos\theta \qquad (3.56b)$$

for potentials $V_o$ and $V_i$ outside and inside the particle respectively, where A, B, C and D are constants of integration which may by determined from the boundary conditions of the problem. Thus, since the potential at the centre of the particle must remain finite, $C = 0$. A long way from the particle, the electric field must be equal to the undistorted applied field $E_a$ so that $B = -E_a$. Finally, at the surface of the sphere, $V_o = V_i$ and the normal component of flux density across the boundary must remain constant (section 2.6.1). These conditions are satisfied when

$$A = \left(\frac{\varepsilon-1}{\varepsilon+2}\right)a^3 E_a \qquad \text{and} \qquad D = -\frac{3E_a}{2+\varepsilon}$$

where a and $\varepsilon$ are respectively the radius and relative permittivity of the particle. Substituting into equation (3.56a) then yields

$$V_o = \left[ \left( \frac{\varepsilon - 1}{\varepsilon + 2} \right) \left( \frac{a}{r} \right)^3 - 1 \right] rE_a \cos\theta \qquad (3.57)$$

for the potential outside the particle. Differentiating and letting $r = a$ gives the field at the particle surface as

$$E_S = \varepsilon' E_a \cos\theta \qquad (3.58)$$

where $\varepsilon' = 3\varepsilon/(\varepsilon+2)$ and ranges from about 1 for a good insulator to 3 for a good conductor. It is seen that the field is greatest at the point marked A in Figure 3.24 where $\theta = 0$. The maximum charge, $q_M$, on the particle is reached when the coulomb field of the charge trapped on the particle surface ($q_M/4\pi\varepsilon_0 a^2$) is equal and opposite to the field at A when charging began. Therefore, the maximum charge on a spherical particle is given by

$$q_M = 4\pi\varepsilon_0 a^2 \left( \frac{3\varepsilon}{\varepsilon + 2} \right) E_a. \qquad (3.59)$$

This equation is known as the Pauthenier limit. For a particle of radius 5 μm and a relative permittivity of 2 in an applied field of 1 MV/m, $q_M = 4.17 \times 10^{-15}$C equivalent to a charge density of $1.33 \times 10^{-5}$ C/m$^2$ on the particle surface.

A similar analysis for a long cylindrical fibre of length L held with its axis perpendicular to the ion flow gives the maximum charge as

$$q_M = 2\pi\varepsilon_0 aL \left( \frac{2\varepsilon}{\varepsilon + 1} \right) E_a. \qquad (3.60)$$

### 3.4.2.2 Charging Time Constant

It is important in the design of electrostatic precipitators to calculate the time required to charge a particle since this will determine the residence time required in the corona for it to become fully charged. The rate of charging is, of course, equal to the number of ionic charges arriving at the surface of the particle per second. We therefore need to know both the density and the velocity of the corona ions close to the particle surface. In the following, it is assumed that (a) the ion density N is equal to that in the uniform, undisturbed part of the field and (b) the ion velocity in a field E is equal to $\mu E$ where $\mu$ is the ion mobility.

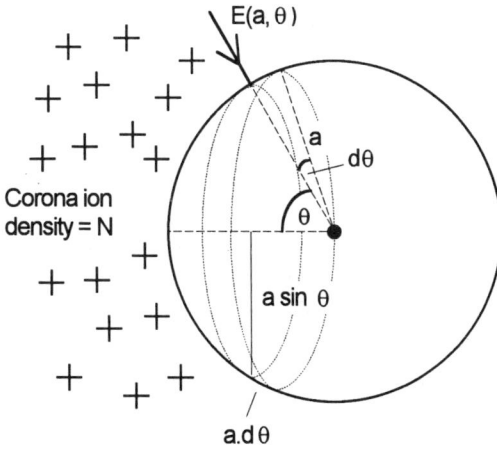

**Figure 3.25**   *Positive corona ions migrate to the particle surface so long as E(a, θ) is directed towards the surface.*

Consider the small annular element of radius $a\sin\theta$ and width $ad\theta$ on the surface of the particle facing the ion source as shown in Figure 3.25. The field $E(a,\theta)$ at this annulus is given by

$$E(a,\theta) = \varepsilon' E_a\cos\theta - \frac{q}{4\pi\varepsilon_0 a^2}. \qquad (3.61)$$

The charge $\Delta q$ arriving in time dt at this surface element is then given by

$$\Delta q = dA.\sigma E(a,\theta).dt$$

where dA is the area of the surface element and $\sigma$ the effective conductivity of the charging zone. Thus

$$\Delta q = 2\pi a^2 \sin\theta d\theta.Ne\mu.\left(\varepsilon' E_a\cos\theta - \frac{q}{4\pi\varepsilon_0 a^2}\right)dt$$

which upon substituting yields

$$\Delta q = 2\pi a^2 Ne\mu\varepsilon' E_a\left(\cos\theta - \frac{q}{q_M}\right)\sin\theta.d\theta.dt$$

where $q_M$ is given by equation (3.59) and q is the total charge on the particle at time t. Integrating over the particle surface gives the total rate of charge accumulation on the particle as

$$\frac{dq}{dt} = \frac{2q_M}{\tau} \int_0^{\theta_c} \left( \cos\theta - \frac{q}{q_M} \right) \sin\theta . d\theta \qquad (3.62)$$

where $\tau = (4\varepsilon_0/Ne\mu)$. At t = 0 the upper limit of the integration is $\pi/2$ since the whole of the front surface of the particle collects charge. However, as charge accumulates on the particle, it can be seen from equation (3.61) that at some critical angle $\theta_C = \cos^{-1}(q/q_M)$ the field reverses so that ions are only collected by the area bounded by $0 < \theta_C$. Equation (3.62) is then easily integrated yielding

$$\frac{dq}{dt} = \frac{q_M}{\tau}\left(1 - \frac{q}{q_M}\right)^2$$

which may be further integrated to give

$$q = q_M\left(\frac{t}{t+\tau}\right). \qquad (3.63)$$

It is now seen that the constant $\tau$ is in fact the time taken for the particle to charge to half its final value. It is in effect a relaxation time for the ionised gas. If the current density, J, and electric field, $E_a$, in the corona are known then $\tau$ is easily calculated from the relation

$$\tau = \frac{4\varepsilon_0}{Ne\mu} = \frac{4\varepsilon_0 E_a}{J} \qquad (3.64)$$

For a corona in which $E_a = 1$ MV/m and $J = 2$ mA/m² then $\tau \sim 18$ ms.

### 3.4.2.3 Bipolar Charging

The analysis given above can be further developed to the case of a bipolar corona where both positive and negative ions are present. From equation (3.61) and Figure 3.25 it is clear that in this situation the particle collects positive ions on the surface bounded by $\theta < \theta_c$ but simultaneously collects negative ions when $\theta_c < \theta < \pi$. Noting that at equilibrium the rates of collection of positive and negative ions are equal, it is readily shown that the limiting charge, $q_{bi}$, in a bipolar ion flow is now given by

$$q_{bi} = q_M \left( \frac{J_+^{\frac{1}{2}} - J_-^{\frac{1}{2}}}{J_+^{\frac{1}{2}} + J_-^{\frac{1}{2}}} \right) \qquad (3.65)$$

where $J_+$ and $J_-$ are respectively the positive and negative ion current densities arriving at the particle and $q_M$ is the limit for monopolar charging.

This is a particularly important result for electrostatic powder coating and precipitator systems because it shows that when back-ionisation is present (section 3.4.2.5) charging efficiency is reduced and the limiting charge on the particle can be much smaller than expected.

### 3.4.2.4 Diffusion Charging

In the above analysis of corona charging it was assumed implicitly that the flow of ions was controlled entirely by the electric field. In other words, the random motion of the ion has been ignored. This assumption only holds for larger particles. For submicron particles, the thermal energy of the ions becomes sufficient to overcome the coulomb repulsion of the charged particle and further charging occurs. Based on a Maxwellian distribution of velocities, the mean translational energy of ions is $3kT/2$, where $k$ is Boltzmann's constant and $T$ the absolute temperature. Diffusion charging will result in a *mean* particle charge $q$ given by

$$\frac{qe}{4\pi\varepsilon_0 a} = \frac{3kT}{2}$$

i.e.

$$q = 6\pi\varepsilon_0 a \left( \frac{kT}{e} \right)$$

where $e$ is the ionic charge. Comparison with equation (3.59) will show that in a field of 1 MV/m and with $\varepsilon = 4$ diffusion charging makes a significant contribution to the charging of particles in the size range 0.01 to 1.0 $\mu$m.

### 3.4.2.5 Back-Ionisation

In electrostatic precipitators, charged particles migrate to the earthed wall of the structure where they "plate" out. If the deposited layer has a volume resistivity $\rho$, and the corona current density arriving at the earthed wall is $J$, then a field $E$ ($= \rho J$) builds up within the layer. When $\rho$ is sufficiently great, $E$ exceeds the breakdown strength of the layer and an electrical discharge is initiated. One consequence of the discharge is that ions of opposite polarity to the corona ions are ejected from the layer and flow towards the corona electrode. This process, known as back-ionisation, reduces precipitator efficiency because it tends to discharge the dust particles. Back-ionisation is also a problem in electrostatic powder coating operations because electrical discharges in the deposited layer can lead to a poor surface finish.

### 3.4.3 Charging of Insulating Sheets

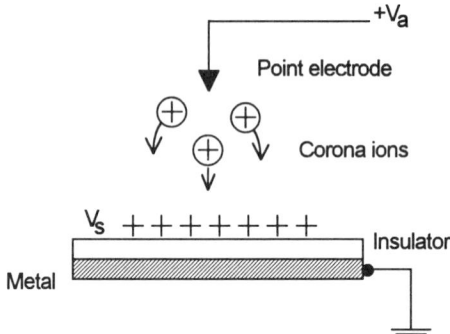

**Figure 3.26**    *Charging insulating sheets using a corona discharge.*

Corona ions are used for charging insulating sheets both in experimental studies of the behaviour of charges on surfaces and in practical applications. The principle is demonstrated in Figure 3.26 which shows an insulating sheet resting on a flat earthed surface in a stream of corona ions from a pointed, high voltage electrode. In this arrangement positive corona ions will flow to the surface of the insulator and become trapped there. The potential of the surface, $V_S$, will increase (equation(2.39)) but ions will continue to flow to the surface until the potential $V_a$-$V_S$ in the air gap becomes less than the corona threshold voltage, $V_C$. The extent to which $V_S$ can be controlled on a long term basis this way is limited, since $V_C$ will depend on environmental factors and on the cleanliness and dimensions of the corona source. And, of course, the danger of overcharging is always present. When the insulator is thin, $V_S$ may rise sufficiently to exceed the breakdown strength of the sheet, leading to electrical breakdown and pinhole formation in it.

Despite such problems, this technique has been widely used to effect permanent polarisation of polymer films,e.g. polyvinylidene fluoride PVDF (for example, see Das Gupta and Doughty, 1978). The method, known as corona poling, involves heating the film above its glass transition temperature, and corona charging it while it cools. Films poled this way are widely used as piezo- and pyro-electric transducers.

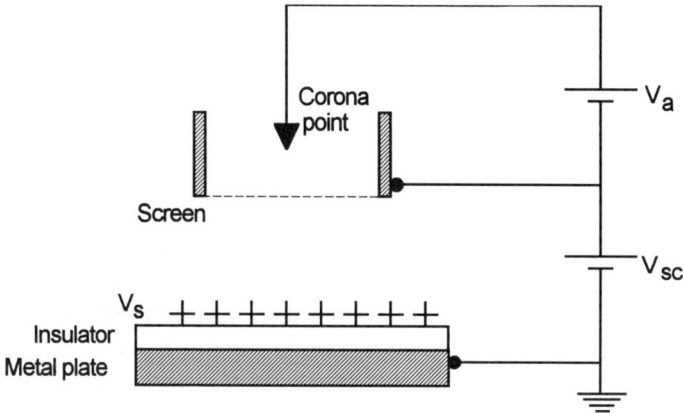

**Figure 3.27**    *Basic construction of the scorotron. The screen voltage $V_{sc}$ controls the final potential $V_s$ on the insulator surface.*

To achieve greater control of the surface potential, a mesh screen electrode can be interposed between the corona source and the surface to be charged (Figure 3.27). Now a stable corona can be established between the corona point or wire and the screen. Corona ions diffuse through the screen and drift under the action of the screen voltage $V_{sc}$ to the insulator surface. Charging will cease when $V_s = V_{sc}$. Such an arrangement is called a scorotron (Schaffert, 1975) and has been widely used in experimental studies of polymer surfaces.

A drawback of the scorotron approach though is the need for two voltage supplies, one to bias the corona point, the other to bias the screen. This is obviated in the corotron (Figure 3.28), the charging device used in electrophotocopiers

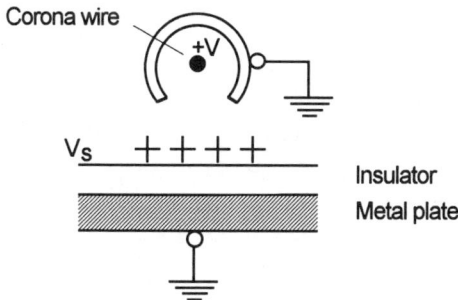

**Figure 3.28**    *The corotron developed for use in electrophotography. The surface potential of the insulator is controlled by the geometry and the exposure time.*

(Schaffert, 1975; Vyverberg, 1965). Here the corona wire is almost completely surrounded by an earthed cylinder. Corona ions diffuse to the insulator surface through a slot in the earthed cylinder. The surface potential now is controlled by fixing the distance between the corotron and the insulator surface and by exposing the surface for a specific time to the corona.

### 3.4.4  Static Eliminators

The discussion above has concentrated solely on the charging ability of a corona. Obviously, corona ions can also be used to neutralise static charge, and both passive and active eliminators are widely used. A passive eliminator is composed simply of an earthed wire or a series of earthed points. If the field set up by a charged sheet is sufficiently great, a corona discharge is initiated at the wire or points. Ions of opposite polarity to those on the charged sheet are drawn out of the ionisation zone and move to the charged sheet where neutralisation is effected.

This is an excellent technique for reducing high concentrations of static charge. However, complete neutralisation is unlikely to be achieved since a minimum charge density must be present on the sheet in order to initiate the corona.

In theory, the residual surface charge can be eliminated by using powered neutralisers which are usually composed of a series of sharp points connected to the output of a 50 Hz, 8 kV step-up transformer. At the peaks of the voltage wave, the source releases sequential bursts of positive and negative ions which are then available for neutralisation. It should be remembered though that 50 Hz eliminators place an upper limit of about 2 m/s on the translational speed of the sheet past the eliminator. At higher speeds, parts of the charged sheet will move through the effective neutralisation zone during the voltage half-cycle when ions of the wrong polarity are being produced.

### 3.4.5 Nature of Corona Ions

From section 3.4.1, it is clear that corona ions are either charged molecular fragments of the gas in which the corona discharge is established or are gas molecules to which electrons from the avalanche process have become attached. In air, the ions normally found in a negative corona are $O_3^-$, $CO_3^-$ and $O^-$, while, in a positive corona, hydrated protons $H^+(H_2O)_n$ seem to predominate with $NO^+(H_2O)_n$ and $NO_2^+(H_2O)_n$ being observed in dry air. Other possible species (Pethig, 1983) include $OH^-(H_2O)_n$, $CO_4^-(H_2O)_n$ and even $O_2^+(H_2O)_n$.

Much of the energy of a corona is dissipated in the form of electrically neutral, though highly reactive, excited species. Consequently, corrosion and erosion are common within corona systems. Not only do corona points become blunt, the counter electrode can suffer severe pitting. Furthermore, any insulation close to the corona source will become charged and may suffer chemical degradation.

## 3.5 OTHER SOURCES OF STATIC

Any ionising radiation such as UV light, X-rays, γ-rays and cosmic rays can charge insulating materials. The main mechanism is the emission of photoelectrons from the insulator surface, which charges positively. This is a particular problem in spacecraft and satellites.

Kilovolt electron beams have been used to charge polymer films to fabricate electrets (Sessler and West, 1975) while radioactive α and β sources have been exploited, with α-sources proving particularly useful as static neutralisers.

## REFERENCES

Bauser H 1974 "Static electrification of organic solids" *Dechema Monograph* **72** 11-28.

Bauser H, Klopffer W and Rabenhorst H 1970 " On the charging mechanism of insulating solids" *Adv Static Electrification* **1** 2-9.

Chapman D 1913 "Theory of electrocapillarity" *Phil Mag* **25** 475-481.

Chowdry A and Westgate CR 1974a "The role of bulk traps in metal insulator charging" *J Phys D: Appl Phys* **7** 713-725.

Chowdry A and Westgate CR 1974b "Comments on contact charging of polymers" *J Phys D: Appl Phys* **7** L149-L151.

CIGRW, Aug. 30 - Sept. 5, 1992, *"Static Electrification in Power Transformers"* Paper presented by Joint Working Group 12/15.13 in the name of Study Committee12 (Transformers) and Study Committee 15 (Insulating Materials)

Cobine JD 1958 *"Gaseous Conductors: Theory and Engineering Applications"* (New York: Dover).

Coste J and Pechery P 1981 "Influence of surface profile in polymer-metal contact charging" *J Electrostatics* **10** 129-136.

Crofts DW 1986 "The static electrification phenomenon in power transformers" *Annual Rept CEIDP IEEE Dielectric and Electrical Insulation Soc* 222-236.

Cross JA 1987 *"Electrostatics: Principles, Problems and Applications"* (Bristol: Adam Hilger).

Das Gupta DK and Doughty K 1978 "Corona charging and the piezoelectric effect in polyvinylidene fluoride" *J Appl Phys* **49** 4601-4603.

Davies DK 1967 "The examination of the electrical properties of insulators by surface charge measurements" *J Sci Instrum* **44** 521-524.

Davies DK 1969 "Charge generation on dielectric surfaces" *J Phys D: Appl Phys* **2,** 1533-1537.

Davies DK 1970 "Charge generation on solids" *Adv Static Electrification,* **1,** 10-21.

Duke CB and Fabish TJ 1976 "Charge induced relaxation in polymers" *Phys Rev Lett* **37** 1057-1078.

Elsdon R and Mitchell FRG 1976 "Contact electrification of polymers" *J Phys D:Appl Phys* **9**

1445-1460.

Fabish TJ and Duke CB 1977 "Molecular charge states and contact charge exchange in polymers" J Appl Phys **48** 4256-4266.

Faraday M 1843 *"Experimental Researches in Electricity"* reprint (New York: Dover 1965).

Finar IL 1973 *"Organic Chemistry"* **Vol 1** 6th Ed (Harlow: Longman).

Gallagher TJ 1975 *"Simple Dielectric Liquids: Mobility, Conduction and Breakdown"* (Oxford: Clarendon).

Gavis J and Koszman I 1961 "Development of charge in low conductivity liquids flowing past surfaces: A theory of the phenomenon in tubes" *J Colloid Sci* **16,** 375-391.

Gibson HW, Bailey FC, Mincer JL and Gunther WHH 1979 "Chemical modification of polymers: XII Control of triboelectric charging properties of polymers by chemical modification" *J Polymer Sci: Polymer Chem* **17** 2961- 2974.

Gibson N 1971 "Static in fluids" *IOP Conf Ser No 11* 71-83.

Gibson N and Lloyd FC 1970a "Effect of contamination on the electrification of toluene flowing in metal pipes" *Chem Engng Sci* **25** 87-95.

Gibson N and Lloyd FC 1970b "Electrification of toluene flowing in large-diameter metal pipes" *J Phys D: App Phys* **3** 563-573.

Gilbert W 1600 *De Magnete* reprint (New York: Dover 1958).

Gouy G 1910 "Constitution of the electric charge at the surface of an electrolyte" *J Physique* **9** 457-468.

Grahame DC 1950 "Effects of dielectric saturation upon the diffuse double-layer and the free energy of hydration of ions" *J Chem Phys* **18** 903-909.

Haenan HTM 1976 "Experimental investigation of the relationship between generation and decay of charges on dielectrics" *J Electrostatics* **2** 151-173.

Harper WR 1967 *"Contact and Frictional Electrification"* (Oxford University Press).

Hays DA 1974 "Contact electrification between mercury and polyethylene: Effect of surface oxidation" *J Chem Phys* **61** 1455-1462.

Henderson D and Blum L 1978 "Some exact results and the application of the mean spherical approximation to charged hard spheres near a charged hard wall" *J Chem Phys* **69** 5441-5449.

Hersh SP and Montgomery DJ 1956 "Static electrification of filaments - theoretical aspects" *Text Res J* **26** 903-913.

Hines RL 1966 Electrostatic atomization and spray painting" *J Appl Phys* **37** 2730-2736.

Klinkenberg A and van der Minne JL 1958 *"Electrostatics in the Petroleum Industry"* (Amsterdam: Elsevier).

Koszman I and Gavis J 1962a "Development of charge in low-conductivity liquids flowing past surfaces: Engineering predictions from the theory developed for tube flow" *Chem Engng Sci* **17** 1013-1022.

Koszman I and Gavis J 1962b "Development of charge in low-conductivity liquids flowing past surfaces: Experimental verification and application of the theory developed for tube flow" *Chem Engng Sci* **17** 1023-1040.

Krupp H 1971 "Physical models of the static electrification of solids" *IOP Conf.Ser. No.11,* 1-16.

Lamo WL and Gallo CF 1974 "Systematic study of the electrical characteristics of the 'Trichel' current pulses from negative needle-to-plane coronas" *J Appl Phys* **45** 103-113.

Lewis TJ 1990 "Charge transport, charge injection and breakdown in polymeric insulators"

*J Phys D: Appl Phys* **23** 1469-1478.

Lowell J 1976 "The electrification of polymers by metals" *J Phys D: Appl Phys* **9** 1571-1585.

Lowell J and Brown A 1988 "Contact electrification of chemically modified surfaces" *J Electrostatics* **21** 69-79.

Lowell J and Rose-Innes AC 1980 "Contact electrification" *Adv Phys* **29** 947-1023.

Montgomery DJ 1959 "Static electrification in solids" *Solid State Phys* **9** 139-197.

Nickell RM and Taylor DM 1991 "Electric fields and potentials in the vicinity of electrostatically charged sheet product suspended between two earthed boundaries" *J Electrostatics* **26** 235-243.

Ohara K 1979 "Contribution of molecular motion of polymers to frictional electrification" *IOP Conf Ser No 48* 257-264.

Outhwaite CW and Bhuiyan LB 1983 "An improved modified Poisson-Boltzmann equation in electrical-double-layer theory" *J Chem Soc: Farad Trans 2* **79** 707-718.

Peek FW 1929 *"Dielectric Phenomena in High Voltage Engineering"* (New York: McGraw-Hill).

Pethig R 1983 "The physical characteristics and control of air ions for biological studies" *J Bioelectricity* **2** 15-35.

Rutgers AJ, de Smet M and de Myer G 1957 "Influence of turbulence upon electrokinetic phenomena" *Trans Farad Soc* **53** 393-396.

Schaffert RM 1975 *"Electrophotography"* (New York: Focal).

Schon G 1965 *"Handbuch der Raumexplosionen"* (Weinheim: Verlag Chemie).

Secker PE 1979 "Static eliminator systems for difficult industrial applications" *Inst. of Phys. Conf.Ser.No. 48*, 115-123.

Sennet P and Olivier JP 1965 "Colloidal dispersions, electrokinetic effects and the concept of zeta potential" *Ind Eng Chem* **57** 32-50.

Sessler GM and West JE 1975 "Electrets formed by low energy electron injection" *J Electrostatics* **1** 111-123.

Shinohara I, Yamamoto F, Anzai H and Endo S 1976 "Chemical structure and electrostatic properties of polymers" *J Electrostatics* **2** 99-110.

Smythe WR 1968 *"Static and Electricity"* (New York: McGraw-Hill).

Stern O 1924 "Theory of the electrical double-layer" *Z Elektrochem* **30** 508-516.

Sze SM 1981 *"Physics of Semiconductor Devices"* 2nd ed (New York: Wiley).

Taylor DM 1974 "Outlet effects in the measurement of streaming currents" *J Phys D: Appl Phys* **7** 394-402.

Taylor DM 1978 "Investigation of charges injected into polystyrene during extrusion" *J Electrostatics* **4** 291-301.

Taylor DM and Lewis TJ 1971 "Electrical conduction in polyethylene terephthalate and polyethylene films" *J Phys D: Appl Phys* **4** 1346-1357.

Taylor DM, Lewis TJ and Williams TPT 1974 "The electrokinetic charging of polymers during capillary extrusion" *J Phys D: Appl Phys* **7** 1756-1772.

Torrie GM, Valleau JP and Patey GN 1982 "Electrical double-layers: II Monte Carlo and HNC studies of image effects" *J Chem Phys* **76** 4615-4622.

Townsend JS 1914 "Potential to maintain current between coaxial cylinders" *Phil Mag* **28** 83-90.

Townsend JS 1915 *"Electricity in Gases"* (Oxford: Clarendon).

Unger BA 1981 "Electrostatic discharge failures of semiconducting devices" *IEEE/Proc IRPS* 193-199.

Van der Meer D 1971 Electrostatic charge generation during washing of tanks with water sprays - I: General introduction" *IOP Conf Ser No 11* 153-157.

Vyverberg RG 1965 in *"Xerography and Related Processes"* Dessauer JH and Clark HE eds (New York: Focal).

Walmsley HL 1982 "The generation of electric currents by the turbulent flow of dielectric liquids: 1. Long pipes" *J Phys D: Appl Phys* **15** 1907-1934.

Walmsley HL and Woodford G 1981 "The polarity of the current generated by the laminar flow of a dielectric liquid" *J Electrostatics* **10** 283-288.

Warburg E 1899 "Point discharges" *Weid Ann* **67** 69-83.

Warburg E 1928 *"Handbuch der Physik"* Vol.14 (Berlin: Springer Verlag).

Zimmer E 1970 "Electrostatic charging of high-polymer insulating materials" *Kunststoffe*, **60,** 465-468.

# CHAPTER FOUR

# MEASUREMENT OF ELECTRIC FIELD

## 4.1 INTRODUCTION

For controlling electrostatic hazard situations and for optimising processes that exploit electrostatic phenomena, it is important to be able to make meaningful measurements which appropriately define the state of the system. As we saw in Chapter 2, electrostatics is essentially concerned with the interaction between charges on the basis of their physical location rather than their motion.Thus charge is the key parameter, but in practice may not be easy to measure directly. Certainly in industrial applications it is more usual to measure the electric field set up by a charge distribution rather than to measure the charge itself. However, it is not merely practical expediency which encourages the measurement of field. Field monitoring has intrinsic merit, in that certain electrostatic phenomena display critical field values for their onset (see Table 4.1).

Particularly in situations where electrostatic charging can cause process disruption, or constitute a hazard, a field meter is usually the first instrument used to provide simple but quantitative monitoring. Thus field measurements made on paper, plastic or textile webs, in process, quickly indicate whether there is a likelihood of sparking, electrostatic "cling", operative shock, or dust attraction. In containers part-filled with electrostatically charged fuel, appropriate field readings enable estimates to be made of the likelihood of an incendive discharge to a nearby earthed object through the flammable vapour above the fuel surface. Finally, field meters are invaluable for checking the correct functioning of processes which exploit electrostatic phenomena - for example electrostatic coating, mineral beneficiation and electrophotography.

129

**TABLE 4.1**    *Electrostatic phenomena associated with critical electric field values (in air).*

| Phenomenon | Electric Field Threshold (kV/m) |
|---|---|
| Sparkover between uniform-field electrodes under laboratory conditions. | 3000 |
| Hair on head "stands on end". Charged web "clings" to earthed roller | 1000 |
| Intermittent sparkover between charged object and ground in a typical industrial situation. | 600 |
| Prickling sensation on hairs on back of hand. Organic-solvent ignition risk. | 200 |
| Onset of corona ionisation at passive neutralising bar. | 40 |
| Dust attraction to charged surface. | 20 |

### 4.1.1 Measurement Philosphy

Although the different field measurement systems to be described in this chapter have very obvious distinguishing features, they nearly all rely on the fact that a conducting plate in effective electrical contact with ground acquires a net surface charge by induction when exposed to an electric field (see section 2.6.2). For a plate of area A the magnitude, $Q$, of the charge induced by an electric field E is given by

$$\frac{Q}{A} = \varepsilon\varepsilon_0 E \tag{4.1}$$

where $\varepsilon \sim 1$ if the plate is in air. It is the flow of these induced charges from ground to the plate surface that is actually monitored in such measurement systems.

## 4.2 THE INDUCTION PROBE

The induction probe is, in concept, the most elementary field meter and directly exploits the charge induction effect discussed above. Suppose the "sensing" plate is connected to ground through a capacitor, C, across which a buffer amplifier is connected, as in Figure 4.1. If charge $-Q$ is induced on the sensing plate, the associated displacement current leaves charges $+Q$ and $-Q$ respectively on the "upper" and "lower" plates of capacitor C, which thus has a voltage V across it such that

$$V = \frac{Q}{C} = \frac{A\varepsilon_0 E}{C}. \qquad (4.2)$$

The meter reading should then be directly proportional to the field.

In practice, induction probes suffer from a number of performance limitations. The finite input impedance and bias current of the amplifier give rise to non-stationary readings and "zero drift" respectively, while an error signal is generated if free charges from the environment (perhaps from a nearby corona source) reach the probe. The limitations due to the amplifier can to a large extent be overcome by suitable device choice and circuit configuration, but the latter effect constitutes a fundamental constraint.

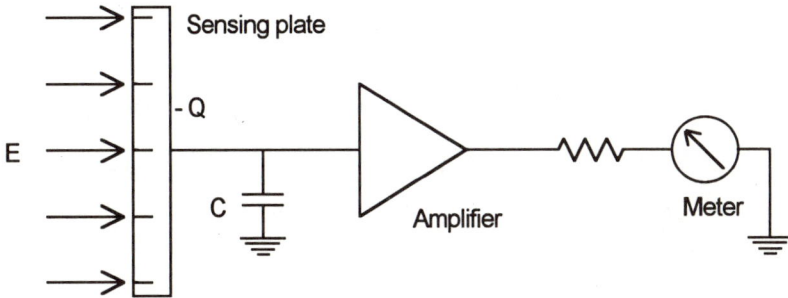

**Figure 4.1**   *Principle of operation of the induction probe field meter.*

Operation of the probe amplifier in virtual earth mode (Figure 4.2), using a feedback link containing the sensitivity-determining capacitor C, means that the negative input terminal of the amplifier remains essentially at earth potential. Thus the sensing plate can be enclosed within an annular guard plate, directly connected to ground, so as to maintain uniform field conditions at the sensing plate.

Suppose that a charge $-Q$ induced on the sensing plate causes a small voltage $\Delta v$ to appear at the negative input terminal of the amplifier. If the amplifier has an open-loop gain of G, then the voltage at the output

$$\Delta v' = -G\Delta v . \qquad (4.3)$$

Since the charge $-Q$ flowed to the sensing plate from the initially uncharged capacitor C then

$$Q = C(\Delta v - \Delta v'). \qquad (4.4)$$

Combining equations (4.3) and (4.4) yields

$$Q = -C\Delta v'(1 + G)/G \qquad (4.5)$$

and

$$E = \frac{Q}{\varepsilon_0 A} \approx -\frac{C}{\varepsilon_0 A}\Delta v' \qquad (4.6)$$

since $G \gg 1$. Thus the meter reading should be directly proportional to the field, the exact relationship between the two being determined by the value of capacitance C.

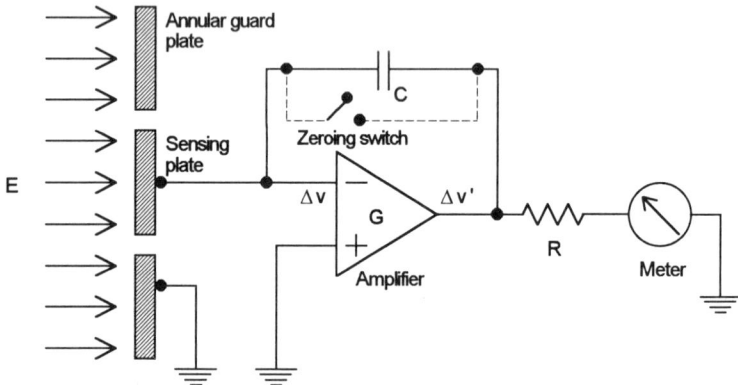

**Figure 4.2**    *Induction probe with virtual earth amplifier.*

The bias current required by the amplifier gives a slow "zero drift" and the resulting error signal, $\Delta v'_e$, observed at the amplifier output must be such as to supply $i_{bias}$ at the amplifier input terminal. Thus

$$\Delta v'_e = \frac{i_{bias} \cdot t}{C} \qquad (4.7)$$

where t is the elapsed time since the capacitor shorting switch was opened (see Figure 4.2).

Suppose a probe of surface area 5 cm$^2$ is required to sense a field of 1000 V/m, and the error due to zero drift is not to exceed 10% after 20 s. From equations (4.6) and (4.7) we may write

$$\frac{\Delta v'_e}{\Delta v'} = \frac{i_{bias} \cdot t}{\varepsilon_0 A E} \qquad (4.8)$$

from which

$$i_{bias} \leq \frac{0.1}{20} \times 8.85 \times 10^{-12} \times 5 \times 10^{-4} \times 10^3$$

i.e. $$i_{bias} \leq 2 \times 10^{-14} \text{ A}!$$

Operational amplifiers with such low bias current performance are expensive, but can result in induction probes of simple design and good performance (Bright et al, 1975).

Blythe and Reddish (1977) describe the use of a matched pair of field effect transistors (FET), connected as a long-tail pair (see Figure 4.3). Careful selection, cleaning and mounting can result in FET gate currents below 1 fA ($10^{-15}$ A), thus resulting in quasi-stable operation.

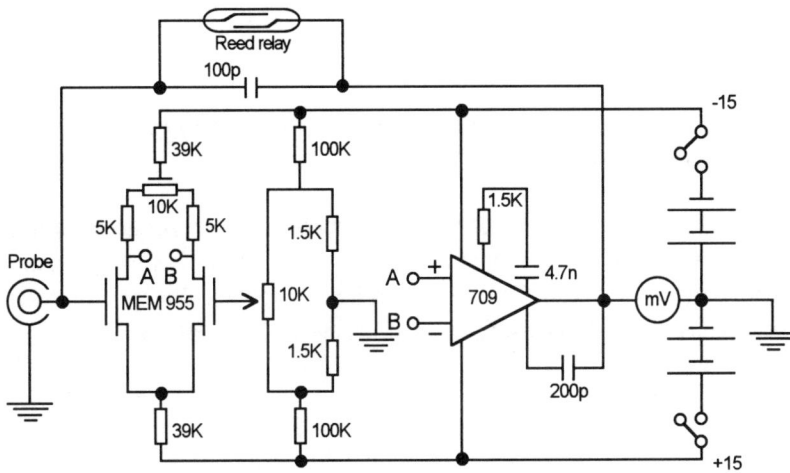

**Figure 4.3**    *Induction probe circuit after Blythe and Reddish (1977).*

An alternative approach (Figure 4.4) is to exploit the low differential bias current exhibited by most FET operational amplifiers. Simple analysis shows the zero-drift/signal ratio is given by the expression

$$\frac{\Delta v'_e}{\Delta v'} = \frac{(i_{bias_2}/C_2) - (i_{bias_1}/C_1)}{Q/C_1} \, t \qquad (4.9)$$

and if $C_1$ and $C_2$ are made equal

$$\frac{\Delta v'_e}{\Delta v'} = \frac{i_{bias_2} - i_{bias_1}}{Q} \, t. \qquad (4.10)$$

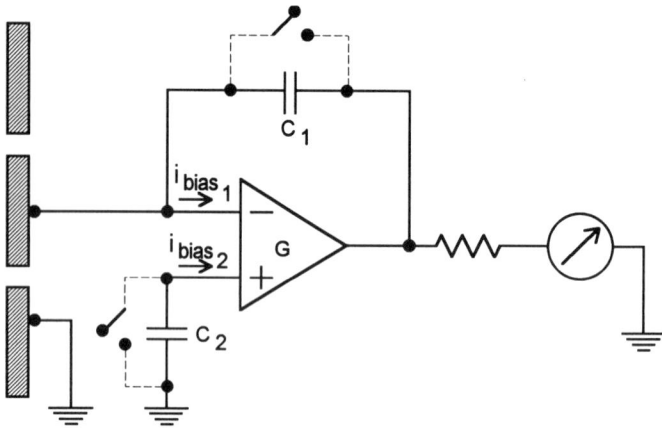

**Figure 4.4**    *Induction probe circuit based on an FET operational amplifier with low differential bias current.*

In this system, two switches of very high open-contact impedance (normally reed relays) are required to temporarily short $C_1$ and $C_2$ just prior to commencing field readings (t=0). Figure 4.5 shows the marked improvement in zero-drift which can be achieved with a low-cost device such as the CA3140 FET op-amp.

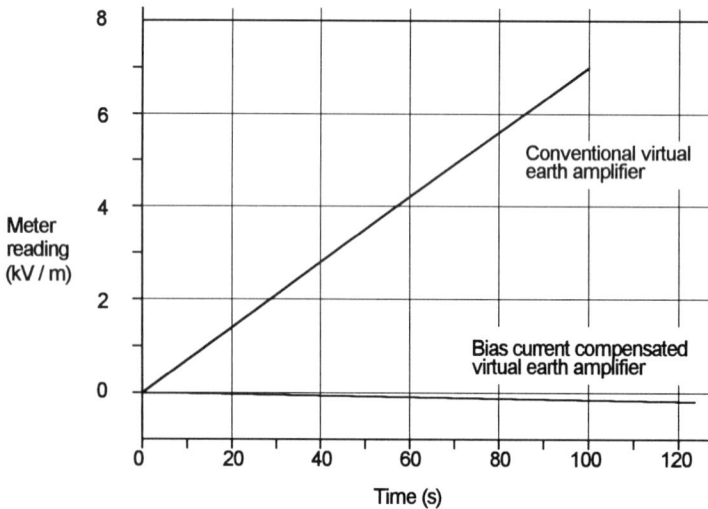

**Figure 4.5**    *Zero-drift behaviour of conventional and bias-current-compensated virtual earth amplifier induction probes.*

The amplifier needs to exhibit very high input impedance ($\geq 10^{12}$ $\Omega$) and also must have a high common-mode rejection ratio. Since the sensing plate voltage gradually changes as t increases, the value of the ground-coupled guard plate in optimising field uniformity is somewhat reduced.

The elegant simplicity of the induction probe is, in practice, marred to a certain extent by the elaborate precautions which are necessary (i) to maintain the high impedance to ground and (ii) to prevent stray leakage currents from flowing to the sensing plate and amplifier input terminals. Typically the sensing plate is mounted on a scrupulously cleaned P.T.F.E. block, while a ground-coupled, screening electrode which completely surrounds the amplifier input terminals is formed using conducting paint.

Prior to making a field reading, the probe charge is reduced to zero by discharging the feedback capacitor C while the probe is in a zero-field environment. It is normally adequate to effect this zeroing merely by screening the sensing plate with a hand. If the sensitivity of the measurement dictates more precise zero-setting, it is doubtful whether an induction probe is the appropriate instrument to use!

The zero-drift tendency of the induction probe is often made less of a practical problem by arranging for an earthed shutter to be moved in front of the sensing plate at the same time as the switch (reed relay) across the feedback capacitor (Figure 4.2) is closed.This re-zeroing operation may either be manually controlled, or automatically initiated every 30 seconds, say.

### 4.2.1 Effect of Atmospheric Currents on Field Measurement

Some estimate of the seriousness of free-charge flow from the environment on induction probe measurements can be gauged from a simple example. Suppose a unipolar charge source (electric air cleaner, powder-coating gun, or ion blower) releases an ion current of 10 µA into the atmosphere, and that the ions disperse isotropically. A probe of area 5 cm$^2$ positioned 10 metres from the source will, theoretically, collect a current $i_p$ given by

$$i_P = \frac{\text{source current} \times \text{probe area}}{\text{surface area of 10 m sphere}} \qquad (4.11)$$

$$= 4 \times 10^{-12} \text{ A}.$$

Only if the current actually collected by the probe is at least 200 times smaller than this theoretical current (this can happen if the ion current is attracted to earthed objects nearer the source) is the error due to the environmental free charge less than the likely zero-drift error (i.e. 10% error after 20 seconds).

## 4.3 THE FIELD MILL

The requirement for a robust field-measuring device, which could operate for long periods without exhibiting significant zero-drift,led to the development of a special type of induction probe in which the sensing plate is repeatedly exposed to and then screened from the incident field by a grounded "shutter". The name "field mill" was coined to describe this new instrument; the brevity and usefulness of the term perhaps compensate for its scientific inaccuracy!

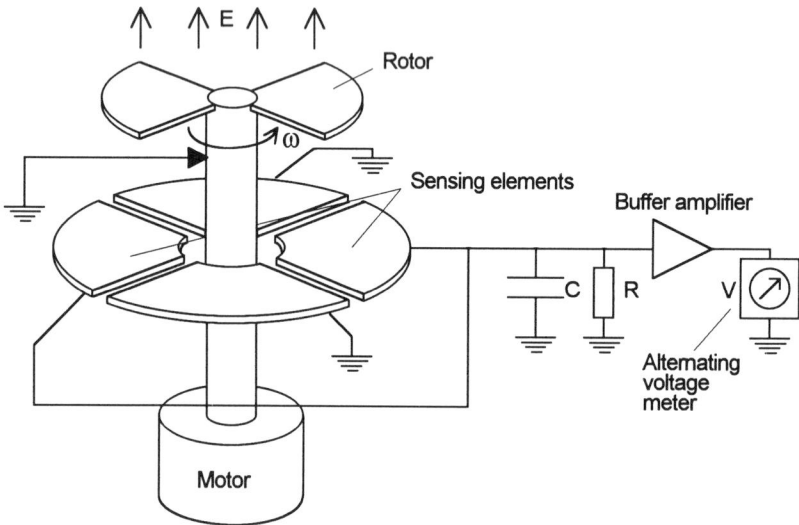

**Figure 4.6**    *Physical configuration of a field mill*

Typically the physical configuration consists of a number of segment-shaped sensing elements, in front of which a similar-shaped grounded rotor turns (see Figure 4.6). The sensing elements are connected to ground via a capacitor C, in parallel with the effective input resistance R of the following buffer-amplifier.

For each complete 360° movement of a 2-blade rotor as in Figure 4.6, the sensing plates are twice exposed to and screened from the incident field. For constant rotor speed (Figure 4.7(a)), the magnitude of the exposed area of the sensing plates varies in triangular fashion (Figure 4.7(b)) and the induced charge must, therefore, have a similar variation (Figure 4.7(c)). In the presence of a non-varying impressed field, any initial direct voltage component across the C/R combination must decay to zero, leaving only the alternating signal created by the field-modulation associated with the rotation of the rotor.

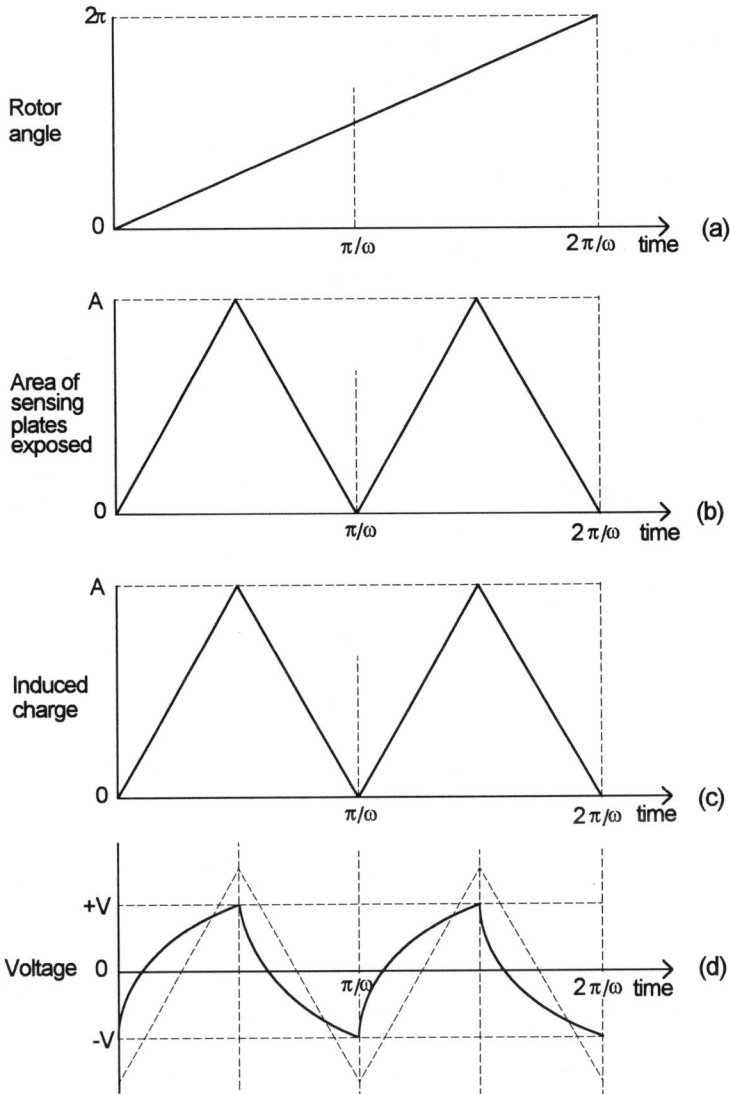

**Figure 4.7**   *Time variation of (a) rotor angle (b) exposed area of sensor plate and (c) the induced charge on the sensor plate for a 2-blade-rotor field mill. The voltage signals at the amplifier input are shown for a purely capacitive impedance (- - -) and a parallel RC (——).*

In the case where R→ ∞, the voltage across C varies linearly with the induced charge, and the voltage/time variation is also triangular (dotted line in Figure 4.7(d)). In general, though, R will be finite and the instantaneous current flow through R results in a more complex voltage variation, but again with a mean value of zero (full line in Figure 4.7(d)).

If the effective area of the sensing plates when fully exposed is A, the rate of change of induced charge can be expressed as

$$\frac{dQ}{dt} = \pm(2\omega/\pi)\varepsilon_0 EA \qquad (4.12)$$

where $\omega$ is the angular rotation frequency of the rotor in radians/s. The induced current flowing to the plate, which changes polarity every quarter cycle of the rotor, must also flow through R and C. Thus

$$\frac{dQ}{dt} = \frac{v}{R} + C\frac{dv}{dt} \qquad (4.13)$$

where v is the instantaneous voltage across R and C. Substituting from equation (4.12) then leads to

$$\frac{dv}{dt} = \frac{(2\omega/\pi)\varepsilon_0 EAR - v}{RC} \qquad (4.14)$$

which may be integrated between the limits $t = 0$, $v = -V$ and $t = \pi/2\omega$, $v = +V$ to yield an expression for the amplitude, V, of the voltage at the amplifier input, i.e.

$$V = \left(\frac{2\omega}{\pi}\right)\varepsilon_0 EAR\left(\frac{(1 - \exp{-\pi/2\omega RC})}{(1 + \exp{-\pi/2\omega RC})}\right). \qquad (4.15)$$

On substituting the series expansions for the exponential terms in equation (4.15) it is readily shown that V is given by

$$V = \left(\frac{\varepsilon_0 EA}{2C}\right)\frac{1 - \pi/4\omega RC + \pi^2/24\omega^2 R^2 C^2 - \dots\dots}{1 - \pi/4\omega RC + \pi^2/\omega^2 R^2 C^2 - \dots\dots}$$

$$= \left(\frac{\varepsilon_0 EA}{2C}\right)\left(1 - \frac{\pi^2}{48\omega^2 R^2 C^2} + \text{higher order terms.}\right) \qquad (4.16)$$

from which we can see that, so long as $\omega^2 R^2 C^2 \gg \pi^2/48$, then

$$V = \varepsilon_0 EA/2C. \qquad (4.17)$$

The peak input signal to the amplifier is directly proportional to the impressed field but independent of rotor speed. This latter characteristic is a very important and necessary feature for battery-powered, portable instruments.

To determine the performance requirements of the buffer amplifier in Figure 4.6 consider an instrument where the capacitor from the sensing-plate to ground has a value of 100 pF. (With smaller values the variation in C due to the variable stray capacitance between the rotor and the sensing plates could introduce errors). If the two-blade rotor turns at 6000 r.p.m. ($\omega = 200\pi$ radians/s), the condition $\omega^2 R^2 C^2 \gg \pi/48$ requires R to be significantly greater than 7.2 M$\Omega$. If $E = 10^4$ V/m and $A = 4$ cm$^2$, the peak voltage across the input C/R will be $+$ 177 mV, an easily measured value. The buffering action is performed by a standard operational amplifier, and provision for switching different capacitors across the input allows measurements to be made over a wide range of field values (typically $10^3$-$10^6$ V/m).

In simple instruments of the type shown in Figure 4.6, the alternating signal from the buffer amplifier is rectified and smoothed, with the following meter calibrated to give the impressed field directly.

### 4.3.1 Detecting Electric Field Polarity

Compared with the induction-probe, the rotor-modulated instrument is far more stable, with zero-drift being negligible over many hours. In the elementary form described so far, however, it cannot discriminate between positive and negative impressed field - a feature which the induction-probe possesses. In fact, as Figure 4.8 shows, the signal appearing across the monitoring capacitor C does contain field polarity information. Upon changing the polarity of the impressed field while maintaining its magnitude a 180° phase shift occurs in the waveform of the input voltage to the amplifier (cf. Figures 4.8 (e) and (f)). This phase change can be exploited to provide readings of both the magnitude and sign of the field. In early instruments (Chalmers, 1953; Mapleson and Whitlock, 1955), complex commutator systems were employed to extract polarity information, but more recently the inclusion of simple phase-sensitive detectors has proved to be more satisfactory.

Figure 4.9 shows conceptually how a typical modern field mill instrument operates. When the chopper vane is correctly positioned, the phase sensitive detector operates as a synchronous rectifier (Macken, 1973), with the output polarity determined by that of the applied field (see Figures 4.8 (g) and (h)).

The optochopper/phase-sensitive detector circuit (Figure 4.10) was analysed by Macken (1973) as a conventional differential amplifier having two modes deter-

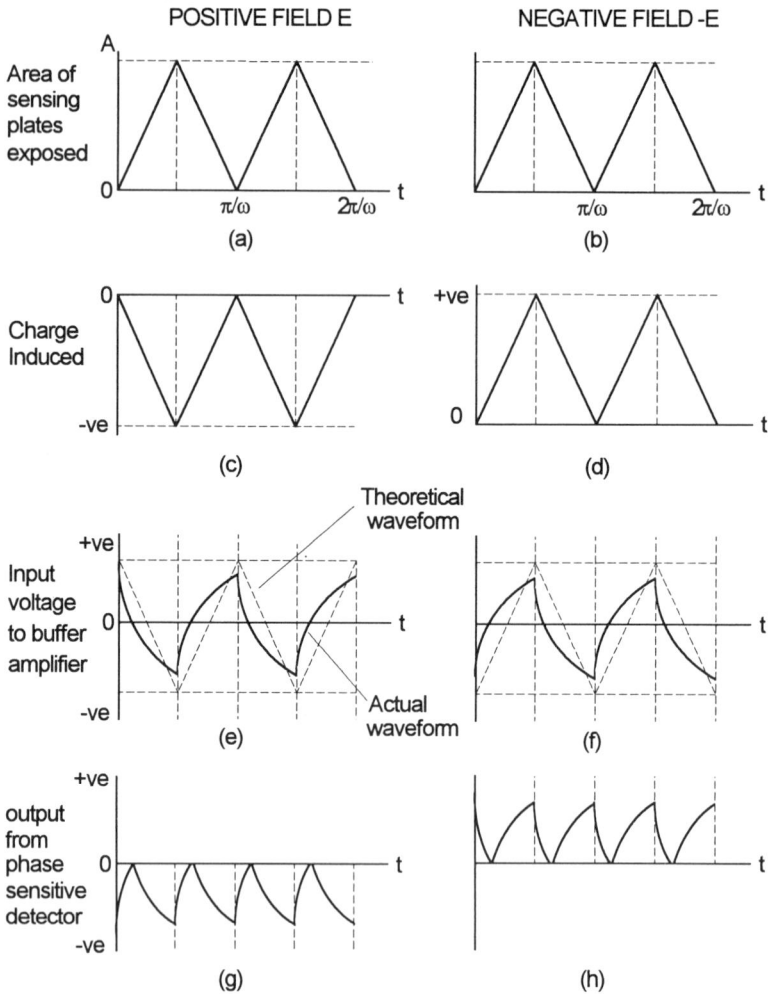

**Figure 4.8**    *The use of synchronous rectification to extract field polarity information.*

mined by the condition of the FET "switch":

(i) With the FET switch open (chopper vane between LED and photo-transistor)

$$v_0 = -v_{in}.$$

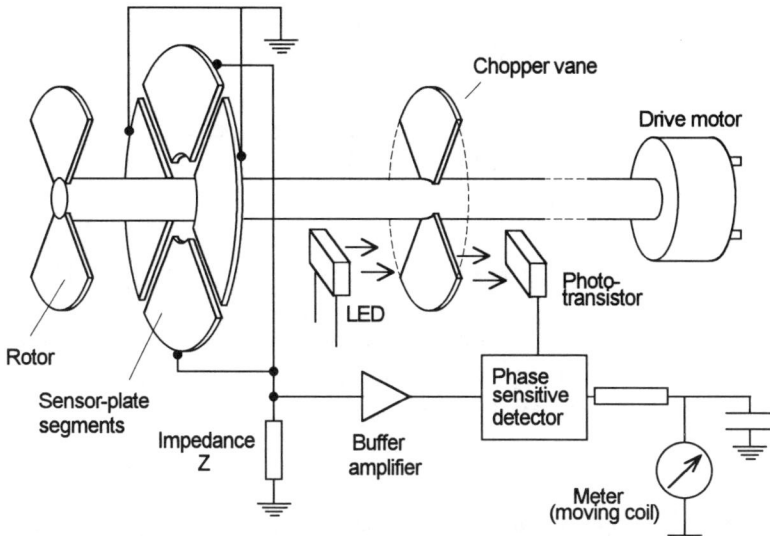

**Figure 4.9** *A modern field mill instrument (IDB Model 107)*

(ii) With the FET switch closed (illumination of phototransistor by LED)

$$v_0 = +v_{in}$$

Thus, adjustment of the angular position of the opto chopper vane relative to the field-modulating rotor is necessary to ensure that phase-sensitive output signal switching takes place when $v_{in} = 0$.

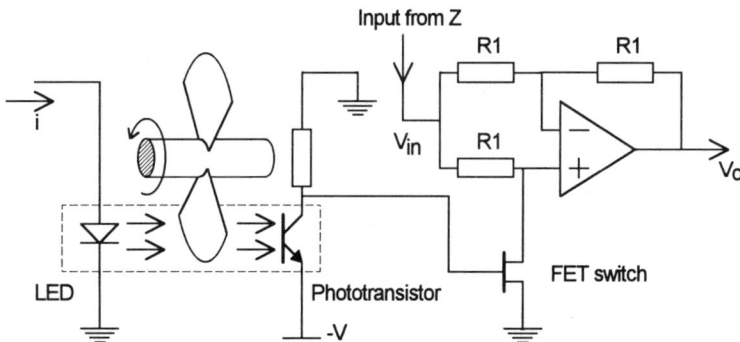

**Figure 4.10** *An optochopper phase-sensitive detector circuit.*

The above analysis shows that any non-varying voltage on the input capacitor of the buffer amplifier (i.e. across impedance Z in Figure 4.9) will be converted by the phase-sensitive-detector into a square wave of equal positive and negative amplitude and thus of mean value equal to zero.

If a field mill such as that shown in Figure 4.9 operates in an environment where it collects free atmospheric charge, the effect is to superimpose an average direct voltage $v_{mean}$ on capacitor C given by

$$v_{mean} = i_a R \qquad (4.18)$$

where $i_a$ is the atmospheric ion current to the sensing plate. As can be seen in Figure 4.11, this direct voltage alters the form of the output signal from the phase-sensitive detector, but the average output voltage $v_{av}$ (related to the shaded areas in Figures 4.11(c) and (d)) is unchanged providing $i_a$ is not so great that the buffer amplifier saturates.

Obviously by displaying the average value of the signal $v_{mean}$ from the buffer amplifier it is possible to monitor the free ion flow to the field meter, enabling

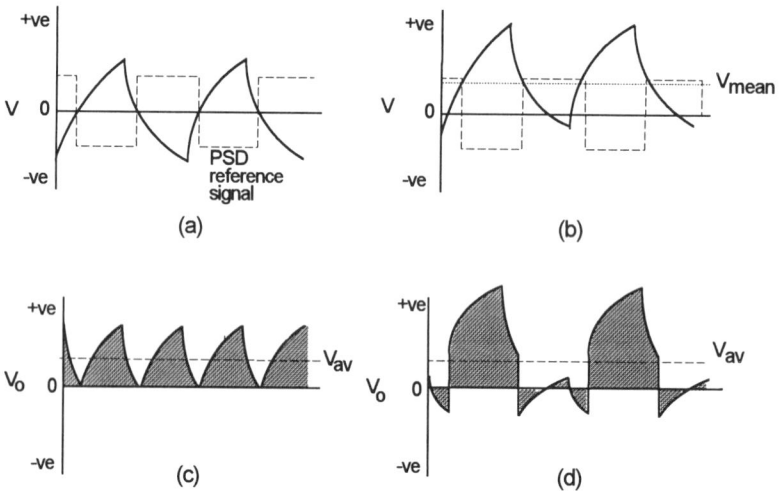

**Figure 4.11**    *(a) Output from buffer amplifier under normal operation and (b) in the presence of an atmospheric current. The corresponding signals $v_0$ at the output of the phase sensitive detector are shown in (c) and (d). The average output signal $v_{av}$ is unchanged by the presence of the direct voltage $v_{mean}$ in (b).*

useful data to be compiled in conditions where space-charge distortion or current flow may be significant.

### 4.3.2  Field Penetration Through Apertures

So far, we have considered induction probes and field mills merely as sensors on which an electric field is impressed. The sensing head of practical instruments is often more complicated than the flat plate considered in our simple modelling up to now.

In many instances the sensing surface is set some distance behind an earthed aperture as shown in Figure 4.12. Such a design (i) minimises the effects of atmospheric ion currents,(ii) screens the probe from extraneous sources of charge and (iii) enables the sensor to be purged when used in a dirty atmosphere. Despite these benefits, the instrument geometry depicted in Figure 4.12 is undesirable for several reasons. Firstly, the fact that the sensing plate is behind an earthed aperture means that the magnitude of the measured field is less than that at the aperture plane. Thus, measurement sensitivity is reduced. Perhaps even more unsatisfactory is the fact that a truly non-uniform field configuration has now been established at the sensing plate. It is therefore more difficult to create simple physical models with which to interpret field readings in relation to the charge distribution in or on the system being assessed.

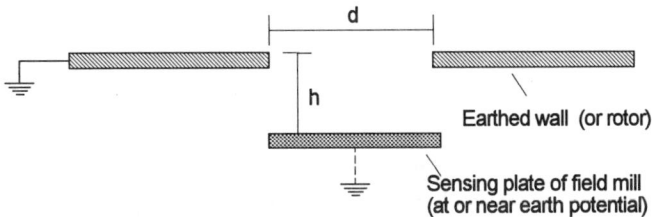

**Figure 4.12**    *Field meter recessed behind an aperture.*

Figure 4.13 shows the typical reductions in field that can be expected in a recessed field meter (Blythe and Reddish, 1977). In this case the measurements were made with an electrode connected to a unipolar high voltage source to establish a field in the measurement zone. Using a conducting electrode, however, gives a more distorted field pattern than would be experienced if a uniformly charged insulator were establishing the field.

It should be noted that, except for small h values, where it is obvious that the monitored field will be more nearly dependent on $(y + h)^{-1}$ rather than $h^{-1}$, the field penetration through the aperture varies only weakly with h. Secker (1977) in

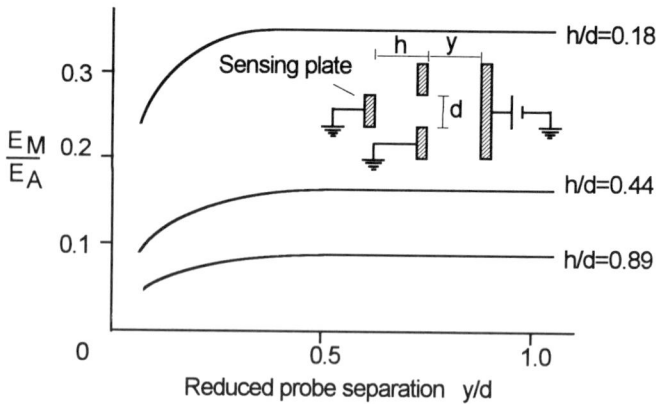

**Figure 4.13**    *Attenuation of recessed field meter. $E_M$ is the measured field when the field at the aperture plane is $E_A$.*

attempting to establish some guidelines useful in instrument design found similar characteristics to those shown in Figure 4.13, but also noted that the shape of the aperture affected the field penetration. In fact a simple empirical relationship was found, namely, that the ratio of the measured field $E_M$ to that which would be found at an earthed surface in the aperture plane $E_A$ is given by

$$\frac{E_M}{E_A} = 0.018 \times \frac{\text{aperture perimeter in mm}}{(h+1)} \tag{4.19}$$

where h is in mm. An effect of the proportionally greater screening as the aperture is reduced in size is that the field at the measuring plane varies approximately as (aperture area)$^{3/2}$ rather than showing a linear relationship as expected. This becomes a serious limitation when using a recessed probe of small area to give good spatial resolution. Sensitivity (detection ability) can then be much poorer than superficial estimates would suggest.

### 4.3.3 Making Field Mill Measurements

Field mills are particularly suitable for measuring the electric fields at the walls of large containers. The meter, which may be considered here merely as an earthed plate, replaces a section of the earthed wall of the containing vessel in order to monitor the field due to the enclosed distributed charge. In such a measurement, the introduction of the field meter does not disturb the field configuration in the container (Figure 4.14).

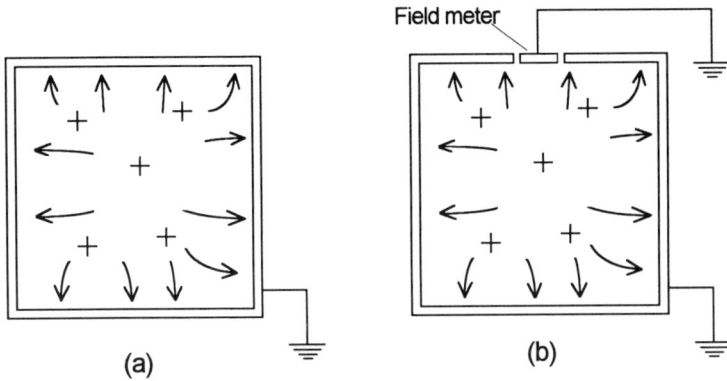

**Figure 4.14**    *The electric field configuration in an enclosed container is not affected by a field meter incorporated into the container wall.*

However, a field measurement without disturbing the system in any way is rarely possible in electrostatics. For example, if charge on an insulating web is being monitored, making the measurement significantly distorts the field pattern. Consider the situation shown in Figure 4.15(a) where a charged web is well removed from any earthed objects. The electric field is directed normally to the surface of the sheet (section 2.3.3) and its magnitude is given by

$$E = \frac{Q/A}{2\varepsilon_0} \tag{4.20}$$

where $Q/A$ is the net charge per unit area of the sheet and includes charges in the bulk of the web as well as on its two surfaces. The factor 2 appears because the field lines emanate equally from either side of the web. When a fieldmeter is brought near to the web, the charge is unaffected but all the field lines are now almost exclusively between the web and the meter, giving rise to a measured field

$$E' = \frac{Q/A}{\varepsilon_0} \tag{4.21}$$

Equation (4.21) does not accurately represent the reading on the field meter since the meter's presence causes the field lines to concentrate at the sensing head leading to an overestimate of the charge density on the web. This problem is eased somewhat by attaching a guard plate to the instrument (Secker, 1975). However, with a meter of the field mill type, there is a further error caused by the rotor and

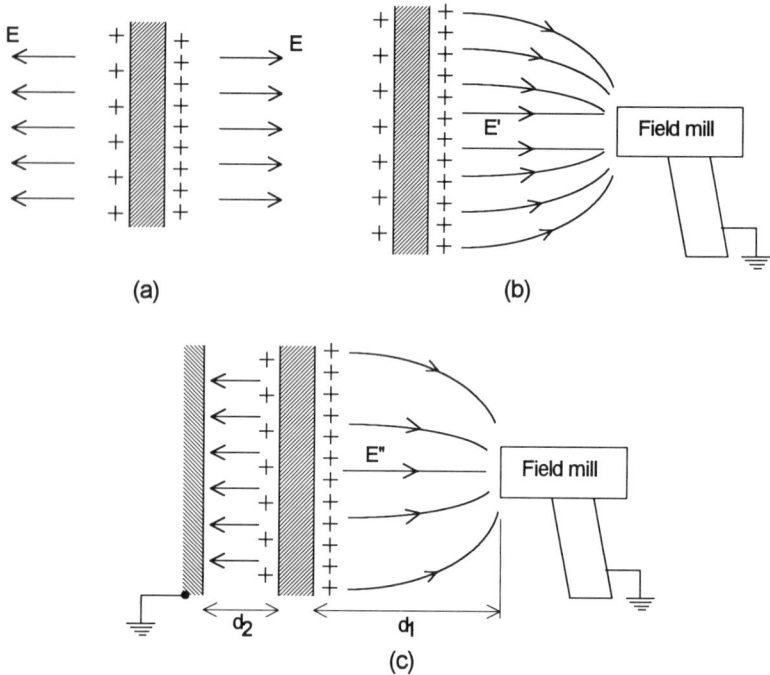

**Figure 4.15**    *Field distribution adjacent to a charged web (a) in free space, (b) near an earthed field mill and (c) between an earthed field mill and an earthed surface.*

sensing plates not being coplanar (section 4.3.2). Consequently, the variation in the sensitivity of the instrument with distance from the web surface is affected both by the degree of macroscopic field uniformity imposed by the guard plate and by the field distortion in the rotor/sensing plates zone, which is a function of rotor blade geometry and the axial separation of the rotor/sensing plates. Figure 4.16 shows the error correction characteristics of two field mills available commercially (Chubb, 1987; Secker, 1975) as a function of the spacing between the meter and a charged web (or, for calibration purposes, a plane electrode raised to high voltage).

Reliable earthing of the rotor tends to be troublesome in instruments used for continuous long-term field-monitoring and where high-speed rotors are necessary to follow accurately any rapid field variations. The difficulty can be overcome by incorporating a second field mill (Figure 4.17) to measure and to compensate for the voltage which may build up on the rotor (Chubb, 1990).

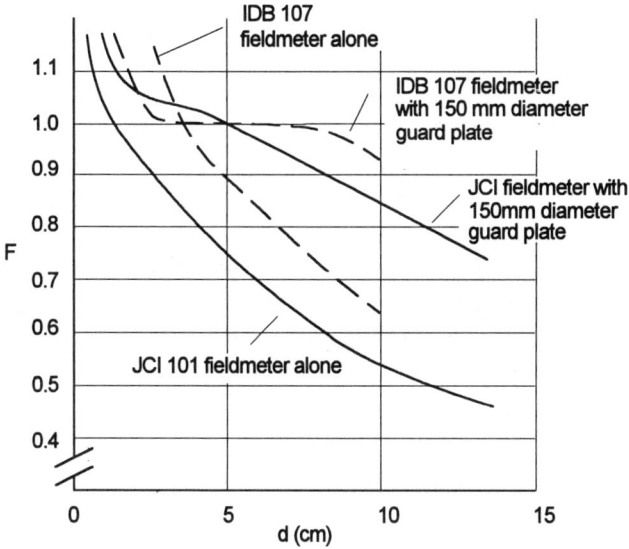

**Figure 4.16**    *Calibration curves for two commercial field mills. F is the correction factor and d the distance from the charged surface to the guard plate.*

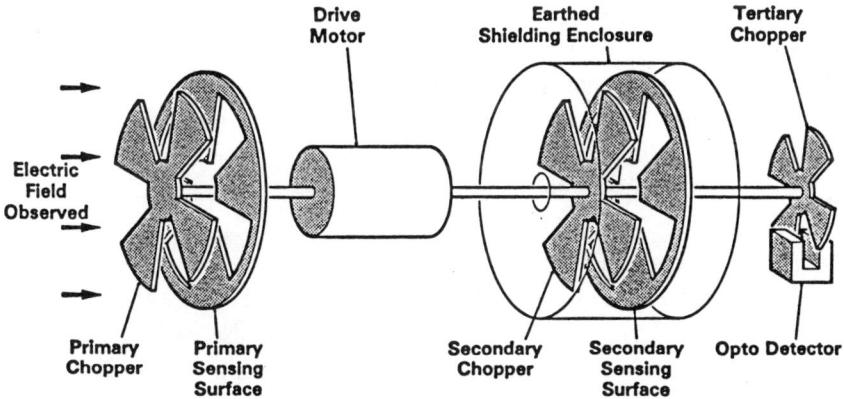

**Figure 4.17**    *A modified field mill in which a second chopper arrangement is used to monitor the potential of the rotor (Chubb, 1990).*

It should be noted that the act of bringing up an earthed meter increases the capacitance-to-earth of the adjacent section of the web. Thus, although the web charge has not changed, its potential decreases (equation 2.38), often by an order of magnitude. As any given section of web moves through process plant its potential will also vary depending on the proximity of nearby earthed surfaces. Web potential is, therefore, not a uniquely defined parameter and for this reason defining the electrostatic state of the web in terms of its potential is inappropriate in such situations.

When a charged web passes close to earthed structures even a field measurement has to be analysed carefully. Figure 4.15(c) shows that some of the field lines from the charged web are diverted from the instrument and couple instead to the earthed surface. It is easily shown that for the simple flat plate geometry considered here the measured field is given by

$$E'' = \frac{Q}{\varepsilon_0 A}\left(\frac{d_2}{d_1 + d_2}\right) \qquad (4.22)$$

where $d_1$ and $d_2$ are, respectively, the distances from the web to the meter and to the earthed surface and are large compared with the web thickness. Clearly, the presence of the earthed surface can cause considerable attenuation of the measured field, particularly when the web is in contact with the plate. Therefore, in a field monitoring activity such as that implied by Figure 4.15(b), which assumes that all the field lines from the charged system end up on the field meter itself, it is important to ensure that the measurement is made at a position where other earthed parts of the process plant cannot materially affect the field distribution. Thus in the typical web/conveyor-belt scenario shown in Figure 4.18, meaningful measurements can be made at positions A, C and E, but not at B or D, where the web charge effectively couples to the earthed rollers.

### 4.3.4  Field Mill Applications

While there are a number of special applications for field mills (see for example sections 5.3, 7.2.1.3, 7.3.3.2 and 7.3.3.3) their principal use is for electrostatic "troubleshooting" - detecting unwanted electrostatic charge accumulation on process plant or product, or monitoring the charge or charged particle deposition efficiency in processes such as electrophotography, electrostatic powder coating, plastic/conductor laminating, etc.

While different field mill instruments have been designed with a range of rotor/sensing-plate sizes and geometries, resulting in internal signal waveforms varying between triangular and sinusoidal, the same signal voltage/ impressed field relationship can be shown to hold (Mapleson and Whitlock, 1955). Many instruments are intended for portable use, and the fact that sensitivity is independent of rotor speed (equation 4.17) if appropriate design constraints have been observed (i.e. $\omega^2 C^2 R^2 \gg 1$) is of great benefit where battery voltage and hence rotor speed are likely to vary.

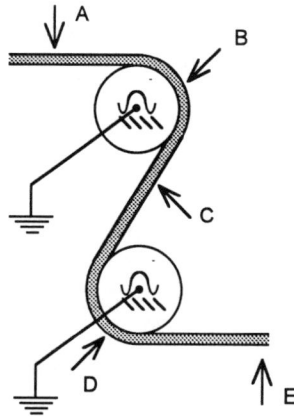

**Figure 4.18**   *Possible sites for measuring the field on a web/roller system. Positions A,C and E are satisfactory. Positions B and D will give significantly attenuated signals.*

In the preceding paragraphs, we have considered the mode of action of the field mill on the basis of a non-varying incident field. Clearly this is an over-restrictive constraint, which in practice can be replaced by the requirement that the change in the incident field during a single cycle of the sensing plate voltage must not be greater than the allowable measurement inaccuracy. Thus by using multi-blade rotor/sensing-plate systems, and maintaining a high rotor speed, it is certainly possible to monitor accurately the rapidly changing field associated with, say, mains-energised (50Hz/ 60Hz) equipment (Secker, 1975b; Waters, 1972).

Waters (1972) used the cylindrical fieldmeter shown in Figure 4.19, operating with an internal frequency of 2 kHz (rotor speed 10,000 rpm), to monitor the surface field on a cylindrical conductor operating below, at and beyond the threshold field for corona onset in a coaxial geometry 50 Hz high-voltage system. The fieldmeter had the same diameter as the central electrode, and could thus be substituted directly for a section of it.

Measurements are sometimes required in environments where the presence of an inflammable gas or vapour (supertanker holds, paper/plastic coating from a

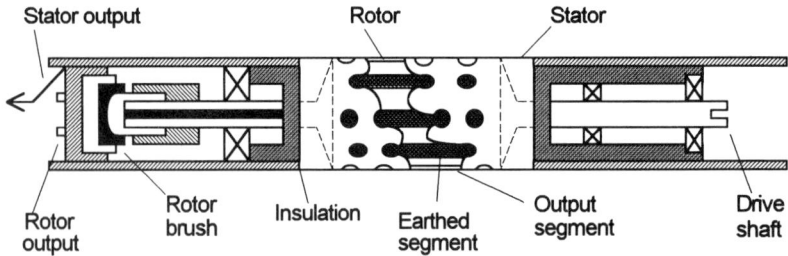

**Figure 4.19**    *Cylindrical fieldmeter used for corona studies.*

high-volatility solvent carrier) demands the use of an intrinsically safe instrument. Specific design rules and operating procedures (BS5501: Parts 1 to 7, 1977; BS 5959: Parts 1 and 2, 1991; HSE Electrical Equipment Certification Guide, April 1992) ensure that, even under component failure conditions, there is no possibility of an incendive discharge occurring within the instrument system. The use of Zener-diode safety barriers (HSE Guide) to supply the signal processing circuits allows the intrinsic safety requirements of this part of the instrument to be easily satisfied. More difficult is the provision of a safe power source to turn the rotor. Novel solutions, involving clockwork drive, air-operated turbines, and epoxy-encapsulated stepper motor/electronic controllers have been exploited.

Figure 4.20 shows an interesting field mill developed by Pollard and Chubb (1975) for operation in dirty and hot environments, and for application in situations requiring intrinsically safe instrumentation. The sensing plates are mounted to give a long surface tracking path to ground, and the maintenance of the required high input resistance condition is continuously monitored by providing a 7 kHz source connected to an electrode viewed by the rotor/sensing-plate combination. The signal from the preamplifier (buffer amplifier) will now contain components due to both the impressed field and the 7 kHz source. These can be separated by active filters to provide the following:

(i) A 7 kHz signal for the reference drive to the phase-sensitive detector, and the calibration output. An appropriate magnitude indicates maintenance of the required leakage inhibition of the sensing plates.

(ii) A low frequency output synchronously rectified by the phase-sensitive detector to give a meter-read signal related to the intensity and polarity of the impressed field.

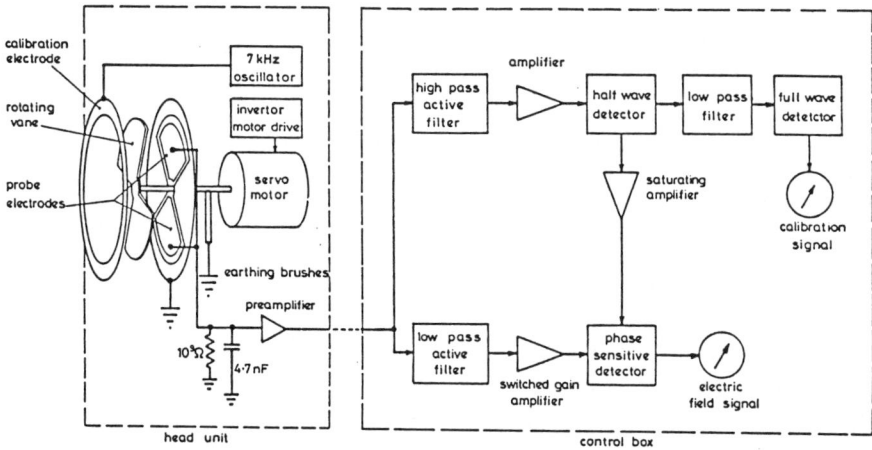

**Figure 4.20** *Special fieldmill developed by Pollard and Chubb (1975) for dirty and hot environments.*

The invertor circuit coupled to the stepper motor, the preamplifier, the 7 kHz oscillator and the ends of the multi-core cable to the control box are all potted in epoxy resin for protection against environmental conditions, and to provide safe operation in flammable atmospheres.

Field mills have been used in helicopters to study the range of field values external to the craft, and to correlate such field values with the onset of radio interference and windscreen discharging (Odam and Forrest, 1969). Rugged construction was necessary to cope with the electrically noisy and vibration-prone conditions, while specifying a logarithmic amplifier made it possible to observe field values between the limits $4 \times 10^2$ and $3 \times 10^5$ V/m, without the need to switch either the input capacitance or the amplifier gain.

In practice, field mills are easy-to-use, undemanding instruments with good field detection sensitivity but poor spatial-resolving characteristics. For consistent readings it is necessary to ensure only that the instrument's battery is charged, the rotor/sensing-plate zone is clean and dry, and that the instrument is connected to earth. It should be noted that when using hand held instruments it should not be assumed that the operator is at earth potential. Insulating footwear results in an ill-defined potential on the operator. Even when wearing static-dissipating foot-wear, the resistance-to-ground through the shoes could be as high as 100 MΩ. So, if measurements are being made close to an ion source such as a corona bar, a leakage current of 10 μA could cause the operator to "float" at a potential of 1 kV.

One final point should be considered when fieldmills are used in high charge situations. Since the instrument is generally earthed to provide a reference potential any sharp edges or corners on or near the sensing head can go into corona which may then neutralise the charge that is being measured, i.e. the meter itself can act as a passive neutraliser!

### 4.4 VIBRATING PROBE FIELD METER

While the field mill has found widespread application as a rugged industrial monitoring instrument, for some applications its size and spatial resolution represent distinct disadvantages. In such cases, the vibrating probe field meter may be applicable. Again the sensor exploits the principle of field-induced charge. Figure 4.21 shows a schematic diagram of the sensor and its associated circuit.

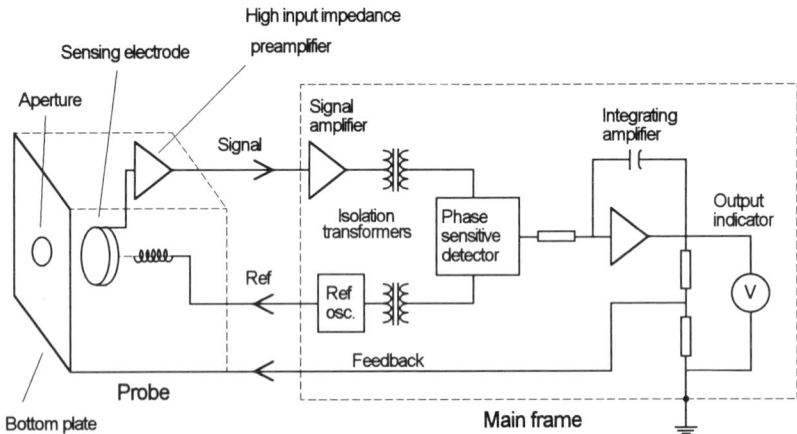

**Figure 4.21**    *Basic elements of a vibrating probe fieldmeter.*

The small sensing plate is located within a screened box having an aperture through which the incident field can penetrate. The sensing plate is mounted on a compliant support, and is caused to vibrate by driving a magnetic coil from a signal generator. Since the field strength decreases as a function of the axial distance from the aperture (see section 4.3.2) vibration of the sensing plate causes a fluctuating potential across the stray or discrete capacitance to ground. A buffer amplifier and phase-sensitive-detector, conceptually identical with those for the field mill, generate a signal dependent on the amplitude of vibration, and the strength and polarity of the incident field. The signal is integrated and inverted, with an appropriate fraction of the resulting direct voltage being fed back to the screening case surrounding the sensing plate. The sensing plate thus sees the sum of two fields - that due to the external incident field, and a field of opposite polarity due to the feedback voltage on the screening case.

Steady state is attained when the net detected field at the sensing plate is zero; the feedback voltage from the integrator is then a measure of the incident field strength. The advantage of this apparently complex null-field detection technique is that vibration amplitude, which is not easy to control precisely over long periods, does not influence the field reading, providing certain essentially geometric design constraints are satisfied (Vosteen, 1974). As with field mills, very stable drift-free operation is obtained.

The sensing head can be made intrinsically safe, and, for use in dirty environments, it is easy to add an air-purging facility to reduce the ingress of particulate matter.

There are a number of other instruments similar in principle to the vibrating probe field meter. These employ either a tuning-probe chopper to generate the varying field at the sensing plate (Vosteen, 1974) or may use a stiff wire which is caused to vibrate in a transverse mode, so repetitively sweeping its sensing face past the field-penetration aperture (Nordhage and Bäckstrom, 1976).

## 4.5 OTHER FIELD DETECTION TECHNIQUES

Field meters based on induced charge effects tend to be used for most field measurements associated with industrial applications. In laboratory investigations, however, a more diverse range of principles have been exploited.

### 4.5.1 The Ballistic Probe
Corbett and Bassett (1971) developed and validated the ballistic probe (see Figure 4.22) first investigated by Pauthenier and Moreau-Hanot (1932). If an isolated conducting spherical particle of small size passes through a space subjected to a field E, and containing free charge, it acquires a limiting charge $Q_0$. Electrostatic repulsion then prevents further free charge reaching the particle. The analysis for this ballistic probe is identical to that presented earlier in section 3.4.2 for the corona charging of particles, so that the limiting charge $Q_0$ in the presence

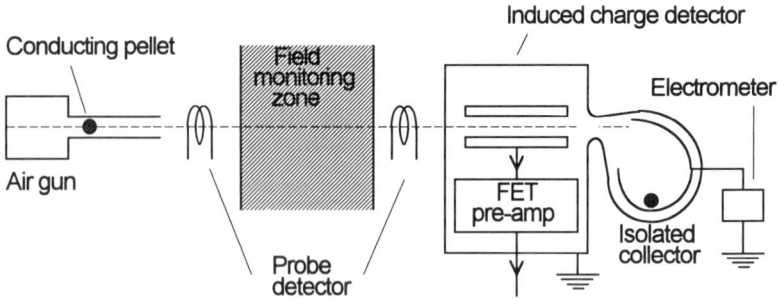

**Figure 4.22**    *Ballistic probe for measuring the strength of an electric field in the presence of free charges.*

of monopolar charges is governed by the Pauthenier limit, namely

$$Q_0 = 12\pi\varepsilon_0 r^2 E \qquad (4.23)$$

where r is the radius of the probe. Charging the probe to this limiting value takes a finite time, so that if it passes through the zone of interest in time t the actual charge, $Q$, acquired is given by

$$Q = Q_0 \frac{t}{t + \tau} \qquad (4.24)$$

where $\tau = 4\varepsilon_0 E/J$ is the effective relaxation time of the ionised gas and J the charging current density. Equation (4.24) can be written in the form

$$\frac{1}{Q} = \frac{1}{Q_0} + \left(\frac{\tau}{Q_0}\right)\frac{1}{t}. \qquad (4.25)$$

If a series of measurements are now made on an unchanging system, but with the probe travelling at different speeds, a range of values are obtained for $Q$ corresponding to different values of t. Then from the intercept of the plot of $1/Q$ versus $1/t$ a value may be deduced for $Q_0$ and hence E may be evaluated. Where the probe travels through a non-uniform field then E is assumed to be the maximum field through which the probe travels.

In practice the probe may be moved through the zone of interest suspended by an insulating thread, or, for more rapid traverses, fired from a pressurised air gun.

The resulting charge on the probe is measured with a non-contact induced charge measuring system (section 6.1).

The ballistic probe has found considerable application in analysing the mechanisms operative at the target surface during electrostatic powder coating, providing valuable evidence to support the hypothesis of "back-ionisation". Additionally, it has been used for evaluating the field distribution in corona-point/plane systems, showing that the magnitude of the field increases adjacent to the plane electrode, probably due to the space charge field of the corona ions but with effects also arising from the charge blocking role of insulating layers on the precipitator surface. The latter phenomenon has considerable significance in optimising the performance of electrostatic precipitators.

### 4.5.2  The Charged Particle Probe

Within fluids, i.e. gases and liquids, particles charged to known levels can be used as field "tracers" to enable the magnitude and direction of the local field to be evaluated.

For a particle of mass m and radius a moving through a fluid of viscosity $\eta$, and subject to both gravity and an electric field E, the acquired velocity is v. The viscous drag force on a particle of radius a is ($6\pi\eta a v$) which, for constant particle velocity, is equal to the local vector sum of the gravitational and electrostatic forces (Stokes' law). Thus

$$6\pi\eta a v = mg + neE$$

where n is the number of units of electronic charge on the particle and m is its mass. Recording the particle movement by photographing it with regularly pulsed illumination enables the particle velocity as a function of position to be calculated, so allowing E to be estimated. This modification of the classical Millikan oil drop experiment has been used to assess the relative importance of gravitational, electrostatic and "ion wind" effects in model studies of electrostatic precipitators (Chubb et al, 1965).

### 4.5.3  Electro-Optical Effects

In certain transparent solids and liquids, the effective refractive index can be noticeably modified by application of an electric field. This effect can be used to modulate the intensity of a light beam, using a physical arrangement such as that shown schematically in Figure 4.23. The light beam enters and leaves the cell through crossed polarisers whose axes are aligned at 45° to the direction in which the field E is applied. The transmitted light intensity, I, varies with the applied field E according to a relationship of the form (Cassidy et al, 1968)

$$\frac{I}{I_m} = \sin^2 \frac{\pi}{2}\left(\frac{E}{E_m}\right)^2 \tag{4.26}$$

where $I_m$ is the maximum light intensity at the detector and $E_m = (2LK)^{-\frac{1}{2}}$ where L is the optical path length over which the field acts and K is the Kerr constant. As the field is ramped the measured light intensity passes through a maximum whenever $(E/E_m)^2$ is an odd integer and a minimum for even integers.

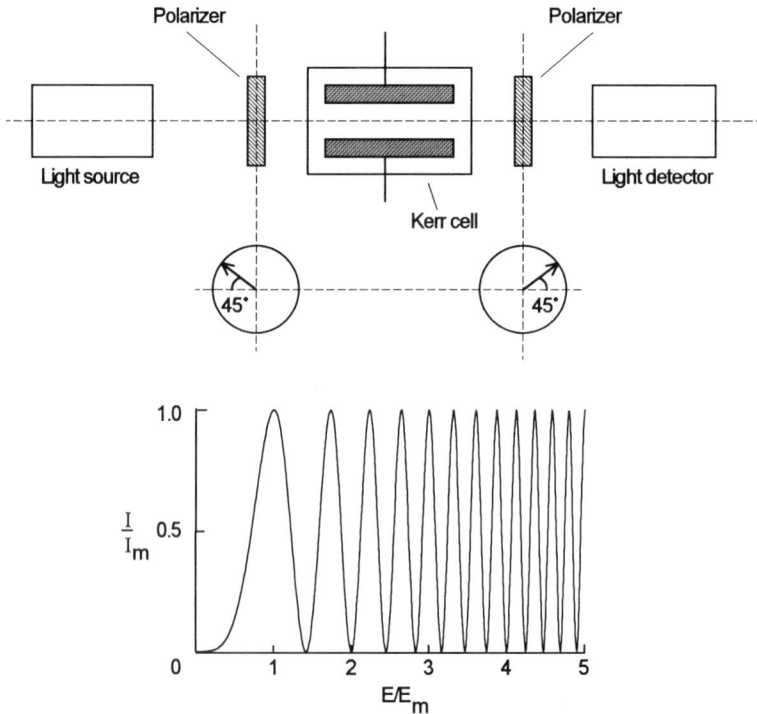

**Figure 4.23**    *Electro-optical arrangement for demonstrating the Kerr effect.*

Exploitation of the phenomenon described by equation (4.26) can, in theory, form the basis of a measurement of field. In practice, applications are based on systems using uniform-field electrode structures. Since the optical response time is very short, rapidly-varying fields can be monitored by means of this technique.

In an alternative application, short-duration light pulses can be created by impressing on the cell a short voltage pulse of appropriate amplitude to make $E=E_m$ in equation (4.26). Such light pulsing devices are known as Kerr (liquid) or Pockel (solid) cells.

## REFERENCES

Blythe AR and Reddish W 1977 "Negative feedback electrometers for electrostatic measurements" *Proc.3rd Int. Congress on Static Electricity, Grenoble* paper 20.

Bright AW, Bloodworth GG, Smith J and Yuratich M 1975 "The development of electronic field meters for use in automatic systems for control of electrostatic charge in fuel tanks" *Proc.Int.Conf. on Lightning and Static Electicity, session 2, paper 4,* Royal Aeronautical Society, London.

Cassidy EC, Cones HN, Wunsch DC and Booker SR 1968 "Calibration of a Kerr cell system for high voltage pulse measurement techniques" *IEEE Trans Intrum Meas* **IM-19** 395-402.

Chalmers JA 1953 "The agrimeter for continuous recording of atmospheric electric field" *J Atmosph Terr Phys* **4** 124-128.

Chubb JN 1987 "Methods and instruments for electrostatic measurements" *Proc Electrostatics Summer School '87* University of Wales Bangor 3.1-3.11.

Chubb JN 1990 "Two new designs of "field mill" fieldmeter not requiring earthing of rotating chopper" *IEEE Trans Industr Appl* **IA-26** 1178-81.

Chubb JN, Bamford WD and Higham JB 1965 "Experimental studies of airborne particle behaviour in corona discharge field" *IEE Colloquium on Electrostatic Precipitation* London.

Corbett RP and Bassett JD 1971 "Electric field measurements in ionic and particulate clouds" *IOP Conf. Ser. No 11,* 307-319.

Macken WJ 1973 "Synchronous and nonsynchronous rectification with op amp" *Electronic Eng* (March) 18-19.

Mapleson WW and Whitlock WS 1955 "Apparatus for the accurate and continuous measurement of the earth's electric field" *J Atmosph Terr Phys* **7** 61-72.

Nordhage F and Bäckstrom G 1976 "Oscillating probe for charge density measurements" *J Electrostatics* **2** 91-95.

Odam GAM and Forrest RH 1969 "An experimental automatic wide range instrument to monitor the electrostatic field at the surface of an aircraft in flight" *Royal Aircraft Establishment, Technical Report 69218.*

Pauthenier MM and Moreau-Hanot M 1932 "Spherical particles in an ionized field" *J Phys Radium* **3** 590-613.

Pollard IE and Chubb JN 1975 "An instrument to measure electric fields under adverse conditions" *IOP Conf Ser No 27* 182-187.

Secker PE 1975a "The design of simple instruments for measurement of charge on insulating surfaces" *J Electrostatics* **1** 27-36.

Secker PE 1975b "The use of field-mill instruments for charge density and voltage measurement" *IOP Conf Ser No 27* 137-181.

Secker PE 1977 "Field perturbations associated with field mill instruments" *Report to the Static Electrification Group* IOP, 5th October.

Vosteen RE 1974 "D.C. electrostatic voltmeters and field meters" *Conf Record 9th Ann Meeting IEEE/IAS* 799-810.

Waters RT 1972 "A cylindrical electrostatic fluxmeter for corona studies" *J Phys E* **5** 475-477.

## CHAPTER FIVE

# MEASUREMENT OF VOLTAGE

## 5.1 INTRODUCTION

In many processes exploiting electrostatic phenomena, a high voltage input is required, either to establish a high electric field, or to generate a strong output of ions or charged particles from a corona source. Since the required electrostatic effect is usually enhanced by increasing the voltage, there is an understandable tendency to operate at as high a voltage as possible. However, theoretical spark-over limits, and rather lower safe, practical operating values mean that in reality the voltage may have to be carefully controlled. Considerations of operator safety, as well as cost, dictate that for many applications the high voltage source has very limited current capability, sometimes augmented by deliberately poor load-regulation characteristics. Voltage measurements, in such circumstances, thus need to be made without drawing any significant current if the behaviour of the system is not to be unduly perturbed by the measurement process.

The particular constraints of low current and high voltage have resulted in a range of different measuring techniques, from which it is usually not too difficult to select one appropriate for any particular application.

## 5.2 VOLTMETERS EXPLOITING ELECTROSTATIC FORCE EFFECTS

From section 2.6.3(b) it may readily be deduced that the force F acting on the image charges induced in a conducting plate of area A, on which a field E is incident, is given by

$$F = \tfrac{1}{2}\varepsilon\varepsilon_0 E^2 A \qquad (5.1)$$

where $\varepsilon$ is the relative permittivity of the medium in which the plate is situated ($\varepsilon \sim 1$ for gases even when pressurised).

159

A variety of voltmeters dependent on this field-related force have been produced over the years, all stemming from Lord Kelvin's "attracted disc electrometer" (Kelvin, 1884). The basic configuration consists of a freely suspended plate or disc, surrounded by an annular guard electrode, with a coaxially mounted counter electrode, also of flat, circular form. If the field between the electrodes is uniform, equation (5.1) can be written as

$$F = \frac{1}{2}\varepsilon\varepsilon_0 A \frac{V^2}{d^2} \qquad (5.2)$$

where d is the inter-electrode spacing. Thus V can be found by measuring F. Unless very special precautions are taken in design, construction and operation, the assumed field uniformity is not maintained, and practical instruments may have a force/voltage relationship of the form

$$F = k.V^x \qquad (5.3)$$

where x is ~2. In such cases calibration by comparison with an absolute measuring device must be performed.

Figure 5.1 shows a relatively crude high voltage meter in which the force-sensing disc in the grounded electrode is mechanically coupled to a pointer moving over a calibrated scale. The maximum field which can be maintained between the electrodes in air at atmospheric pressure ($\sim 1.5 \times 10^6$ V/m) gives a force on a 5 cm diameter disc of only 0.04 N ($\sim 4$ g). Clearly, the moving plate and coupled meter point must be very delicately suspended if friction effects are not to perturb the voltage measurement, but the instrument then becomes very susceptible to mechanical damage.  Since the force on the attracted plate and hence the deflection of the meter pointer vary approximately as $V^2$, the useful measurement range for a given inter-electrode spacing, d, is extremely limited, amounting at best to a 3:1 variation (say 100 kV down to 35 kV). Moreover, the fact that the attracted plate must move significantly (1 or 2 mm) to give the necessary pointer deflection means that in use it does not lie accurately in the surface plane of the earthed electrode. The associated distortion of the electric field means that if d is changed the instrument has to be recalibrated to maintain reasonable accuracy. Despite its limitations the attracted-plate voltmeter has the great advantage of requiring only an initial small charging current when connected to a direct voltage source (the capacitance of the H.T. electrode to ground is only a few picofarad), and thereafter not loading the supply at all.

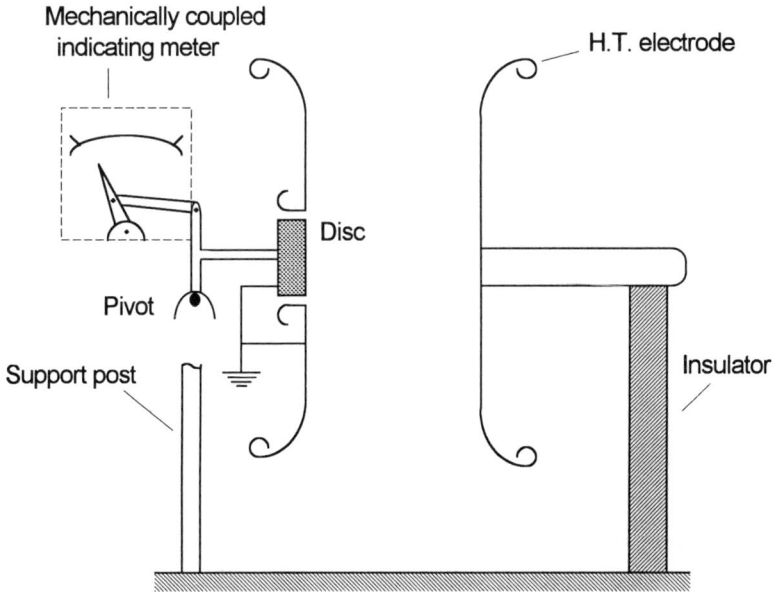

**Figure 5.1**    *Simple attracted-plate electrostatic voltmeter*

As will be obvious from equation (5.2), the attracted-plate voltmeter has the capability of being an absolute measuring instrument. Careful profiling of the electrodes, choice of appropriate relative dimensions for the attracted plate and its surrounding guard electrodes, and maintenance of the coplanarity of the plate/guard electrode surfaces ensure that a uniform field is established, so enabling the applied voltage to be "weighed".

High-precision-attracted plate voltmeters are bulky and expensive structures and are thus the provenance of national standards laboratories. Figure 5.2 shows in outline the instrument constructed at the National Bureau of Standard laboratory (Brooks et al, 1938). It can make measurements accurate to within a few parts in $10^4$ for alternating voltages up to 275 kV r.m.s. (Since the force of attraction varies as $V^2$, measurement of the mean force gives the r.m.s. voltage directly). The high voltage stress-relieving dome incorporates the attracted plate and weighing balance, with a mirror/light beam system for detecting the exact balance point when the surface of the attracted plate and guard electrode are co-planar. Voltage grading hoops connected to a capacitance-divider chain maintain the field uniformity over the 100 cm-long inter-electrode gap.

To avoid problems of draughts, dust and long inter-electrode gaps other absolute voltmeters have been operated in pressure chambers, Böcker (1939) using carbon dioxide at 15 atmospheres for a 600 kV instrument and Bowdler (1955) employing nitrogen at 15 atmospheres or a mixture of nitrogen and arcton-6 (dichlorodifluoromethane) at about 4 atmospheres for measurements up to 350 kV peak (a stress withstand capability of about 10 MV/m).

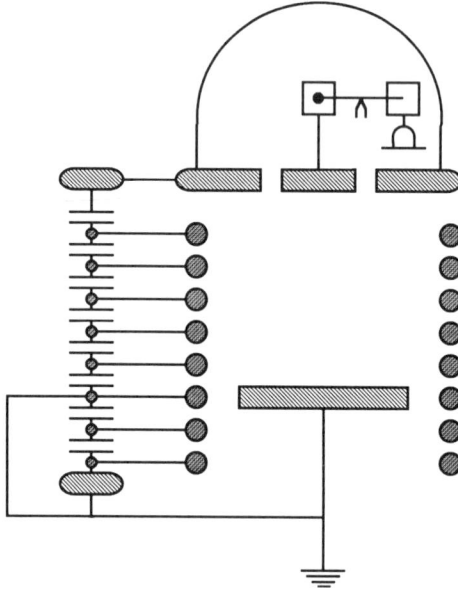

**Figure 5.2**   *The absolute attracted-plate voltmeter by Brooks et al (1938).*

In Bowdler's instrument, built at the National Physical Laboratory, the attracted plate is incorporated into the ground-connected lower electrode. The force exerted by the electric field on the attracted plate can be compensated either by added weights, or by means of a current passed through a coil located in the annular gap between the pole pieces of a permanent magnet. Measurement accuracy was shown to be within 0.1% above 30 kV and about 1% at 1 kV.

A commercially produced voltmeter (Industrial Development Bangor (UCNW) Ltd.) very similar in concept to Bowdler's instrument was developed for operation over the direct voltage range 3 kV to 170 kV. The physical construction of the instrument is shown in Figure 5.3, while Figure 5.4 shows its electronic circuits in schematic form. Miniaturisation compared with standards laboratory instruments has been possible by careful attention to electrode design, the incorporation of automatic electromagnetic compensation, the use of solid state control circuits and a slight relaxation of accuracy standards to the limit normally specified for industrial grade instruments.

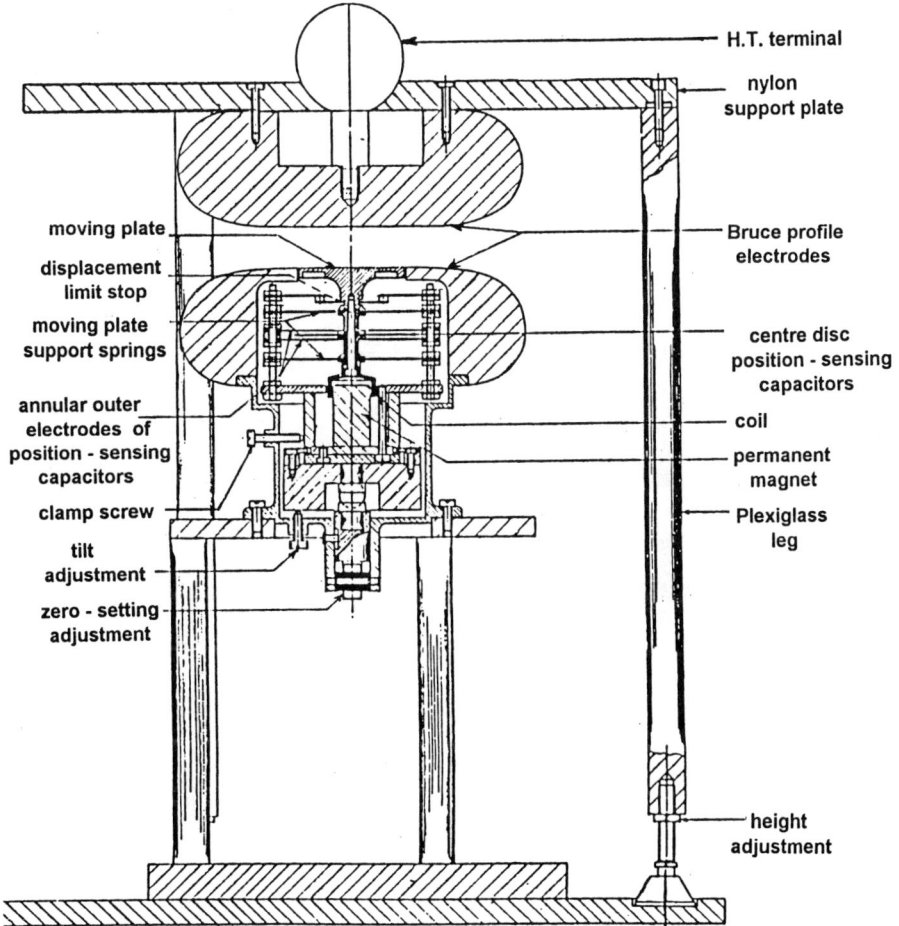

**Figure 5.3**    *Section through electrode/moving-plate assembly of air-operated attracted-plate voltmeter.*

To provide a compact instrument, electrodes of Bruce profile (Bruce, 1947a) were specified since a uniform field over the central portion of the opposing surfaces is maintained for a wide range of gap spacings, even when the electrodes are slightly misaligned or transversely displaced with respect to each other.

The attracted-plate system set into the flat portion of the earthed electrode is suspended on two light, phosphor-bronze springs. Mechanically coupled to the attracted plate is the disc electrode which forms, with the adjacent fixed plates, the position transducer capacitors $C_1$ and $C_2$. The feedback coil by which an electro-magnetic force can be applied to counteract the electrostatic force is also attached to the attracted plate. The complete system and its suspension constitute a single unit which can be accurately positioned within the earthed Bruce profile electrode.

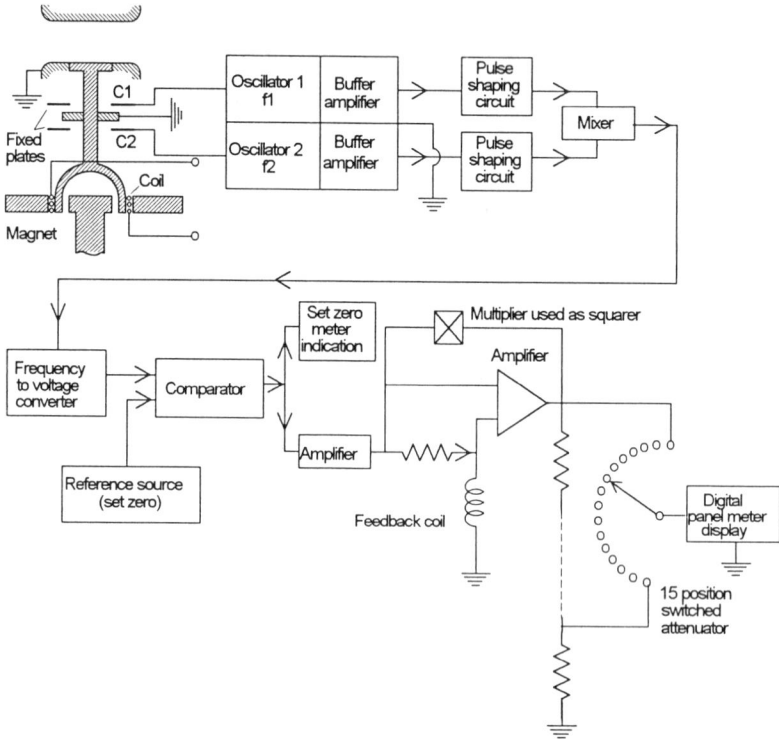

**Figure 5.4**    *Electronic system schematic for attracted-plate voltmeter.*

The capacitors $C_1$ and $C_2$ constitute the frequency-determining elements of Wien bridge oscillators, the outputs from which are mixed and smoothed to provide a difference frequency signal. This is then passed to a frequency-to-voltage convertor to yield an output voltage sensitively dependent on the displacement of the attracted plate. A buffered and amplified version of the displacement signal is used to pass a current through the coil of the electromagnetic compensation-force system.

A voltage signal proportional to the coil current is fed to an amplifier with a squaring multiplier element incorporated in the feedback loop. The output signal is thus proportional to (coil current)$^{1/2}$ and, hence, directly proportional to the applied voltage. In order to effectively cover the 3 to 170 kV operating range, the inter-electrode gap of the voltmeter can be increased in 0.5 cm increments from 1 cm to 8 cm by placing accurately dimensioned aluminium spacer pieces under the H.T. electrode support legs. The gap is "dialled-in" on a multi-step attenuator incorporated in the input circuit of the digital display meter which provides a direct readout of voltage.

For alternating voltages with frequencies greater than 5 Hz the display shows the r.m.s. value, due to the damping provided by the suspension of the attracted plate. Measurement inaccuracies at any given gap spacing are less than 2% of the maximum voltage at that gap for both direct and alternating voltages.

Waterton (1976) describes an electrostatic-attraction voltmeter in which an earthed, hemi-spherical shell mounted eccentrically on a vertical shaft is influenced by a coaxial high voltage electrode consisting of a hemisphere mounted on a cylinder. Paired springs control the rotation of the attracted hemisphere support shaft, and a mirror on the shaft, in conjunction with a point light source and an extended translucent scale, provide a sensitive means for displaying the voltage read-out. By filling the instrument with sulphur hexafluoride at 4.8 atmospheres, average working stresses up to 10 MV/m can be used, giving large deflecting forces and higher accuracy. Measurements have been made from 50 kV to 300 kV accurate to within 0.3% of full scale deflection for both direct and alternating voltages. As with many precision attraction-type voltmeters, oscillation problems were initially experienced with alternating voltages and it was found necessary to increase the magnetic damping (provided by a vane mounted between permanent-magnet pole-pieces), eventually making the movement system effectively aperiodic.

A further group of attraction-based voltage-measuring instruments is typified by the Bruce ellipsoid voltmeter (Bruce, 1947b). A conducting, lightweight ellipsoid (nominal axes 40 mm and 6 mm) on a flexible insulating thread is set to swing normally to the common axis of the uniform field electrodes (diameter of uniform field region >1 m). Application of voltage alters the oscillation frequency, resulting in a voltage/frequency relation:

$$V = C\left(f_V^2 - f_0^2\right)^{\frac{1}{2}} \tag{5.4}$$

where $f_V$ is the frequency with voltage V applied, $f_0$ is the frequency with no voltage applied and C is a constant determined by the ellipsoid shape. It is essential that the ellipsoid does not acquire a net charge, so care has to be taken to ensure corona-free operation up to the maximum working voltage. For Bruce's instrument, accuracy to within 0.03% was claimed up to a maximum voltage of 250 kV (r.m.s.).

A variety of electrode systems and modes of oscillation modes have been exploited for oscillation-type voltmeters (reviewed by Bradshaw et al, 1948), but the technique suffers from the generic disadvantage of requiring the voltage to remain steady for a significant time while the oscillation frequency is accurately measured. Moreover, environmental conditions, e.g. draughts, air-borne dust etc, can result in reading inaccuracies. This type of instrument is, therefore, only really suitable for the specialised research laboratory.

One of the commonest type of electrostatic force voltmeters is the interleaving vane voltmeter, shown in outline in Figure 5.5. A voltage V is applied between the fixed and moving vanes. This causes the suspension to rotate through some angle $\theta$ at which the restoring torque T balances the electrostatic force. If, with this degree of vane overlap, the capacitance between the fixed and moving vanes is C, then the stored energy, U, in the system is

$$U = \tfrac{1}{2}CV^2. \tag{5.5}$$

Assume now that the electrostatic force is increased by a small amount causing a further rotation $\Delta\theta$ of the suspension. The work done on the vanes must equal the change in stored energy $\Delta U$. In the limit where $\Delta\theta$ is considered infinitely small

$$T = \frac{dU}{d\theta} = \frac{1}{2}V^2\frac{dC}{d\theta}. \tag{5.6}$$

If there are n "double-wing" moving vanes, separated on each side by a distance d from fixed vanes, then

$$\frac{dC}{d\theta} = \left(\frac{4n\varepsilon_0}{d}\right)\left(r_1^2 - r_2^2\right)$$

yielding finally

$$T = \left(\frac{2n\varepsilon_0}{d}\right)\left(r_1^2 - r_2^2\right)V^2. \tag{5.7}$$

Choice of intervane spacing d, and the number of moving vanes n, determines the maximum operating voltage and sensitivity of the instrument. Designs exist for meters with full scale values ranging from 100 V to 10 kV. The disadvantages of limited effective scale range (deflection proportional to $V^2$), and rather delicate suspension, are offset by the ability to read r.m.s. values, and the relative ease with which extremely high input impedance (> $10^{11}$ Ω) can be maintained.

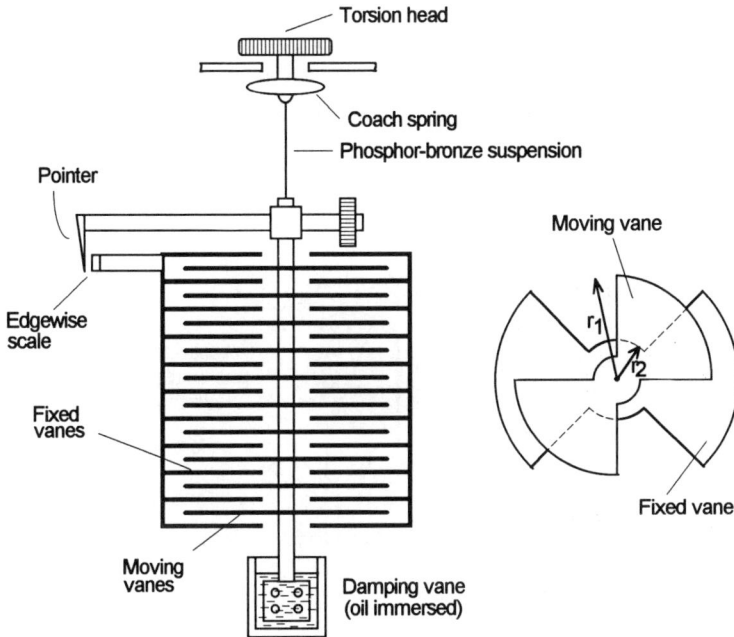

**Figure 5.5**    *Interleaving vane voltmeter.*

## 5.3 FIELD MILL VOLTMETER

The availability of field mills as standard monitoring instruments has provided a simple means of realising a high input-impedance voltage measurement system. Initial use of such systems was for providing output control of Van de Graaff generators (Trump et al, 1940), in which a field mill (normally then called a generating voltmeter) set into the earthed wall of the pressure vessel sensed the field due to the voltage (up to several MV) on the EHT terminal.

Figure 5.6 shows a commercially available field mill voltmeter, in which a field mill mounted in an earthed electrode located on a common axis with the high voltage electrode is enclosed in a pressure vessel to allow high operating stresses and preclude variations due to environmental conditions.

The instrument illustrated is able to operate at peak voltages up to 300 kV. Simpler systems for operation in air at voltages up to 150 kV have been created merely by fitting a standard field mill into the earthed electrode of a Bruce profile pair, set at an appropriate gap spacing (10 cm). Provided the inter-electrode gap is maintained constant at the spacing at which the instrument was calibrated, it is of no concern that the field across the inter-electrode space is not strictly uniform. In practice, the constraint of fixed-gap operation does not constitute a serious limitation, since range changing, using a switched gain amplifier, or switched input capacitors, enables measurements over a 1000:1 voltage variation to be effected with relative ease.

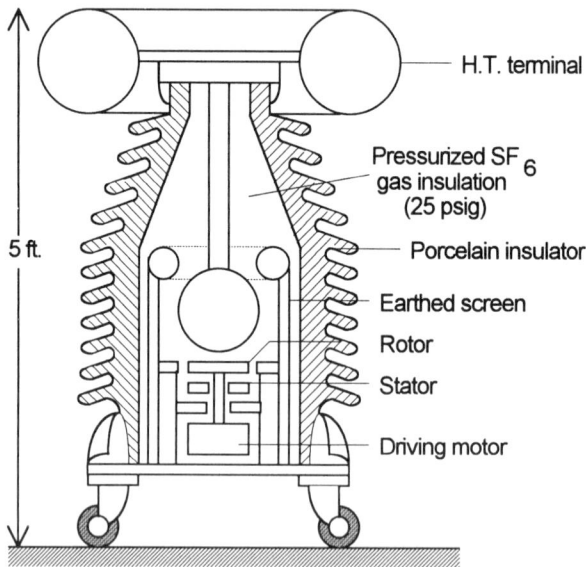

**Figure 5.6**   *Pressurised field mill voltmeter (Ion Physics Corporation, Burlington, Massachusetts).*

Operation of the field mill voltmeters discussed so far presupposes that the voltage output from the source being monitored is invariant, or at least only varies very slowly. However, by increasing the rotor speed, or using a multiblade rotor/sensing-plate configuration, it is possible to generate an output from the field mill which is an accurate, attenuated reproduction of the waveform appearing on the high voltage electrode (Trump et al, 1940). Measurement of a 50 Hz waveform requires a field modulation at the sensing plate giving an internal frequency in the field mill circuits of some 2 kHz, if important low-order harmonics are to be accurately monitored. Where combined direct and alternating waveforms are present (e.g. in the unsmoothed output from a step-up transformer/ rectifier circuit), the addition of selective filter circuits to the field mill output, and standard electronic circuit modules, enable the mean direct voltage, r.m.s. voltage, peak voltage and third harmonic ripple amplitude to be individually displayed.

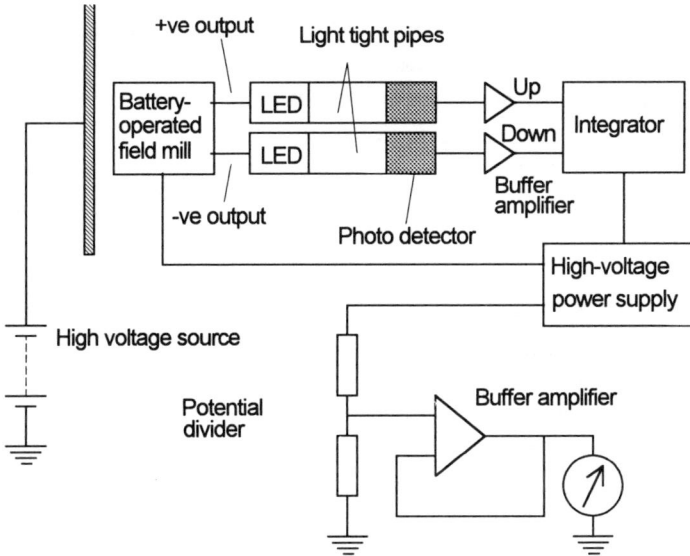

**Figure 5.7**    *Non-contact field mill voltmeter.*

The system shown in outline in Figure 5.7 is an application of the field mill transducer where direct connection to the high voltage source is not  possible or practically undesirable. The field mill is battery operated, and the polarity of the output is transmitted to the ground-connected control equipment by means of LED/light-pipe/phototransistor links capable of withstanding some 30 kV. Effect-ively the polarity signal from the field mill results in a raising or lowering of the output voltage from the instrument's own EHT supply, until the field mill as a whole is at the same potential as the source being monitored in which case the fieldmeter "sees" a zero field.

The EHT supply of the instrument has sufficient output power to drive an associated potential divider so that the floating voltage of the field mill, and hence the source potential,can be directly displayed. Measurement accuracy is to within $\pm$ 50 V over the whole operating range up to 25 kV (Secker, 1975).

If the conductor on which the source potential is being measured is small relative to the dimensions of the sensing head of the field mill, the field mill will settle at a "null" value corresponding to a potential between that on the conductor and earth. The exact value will be determined both by the size and potential of the high voltage source, and the disposition of earthed objects within the "field of view" of the field mill.

Making the voltage-sensing head significantly smaller than that of the typical field mill contributes both to measurement accuracy, and to the ease of positioning the transducer at an optimum measurement location. Figure 5.8 is an outline of a commercially-available modulated-field transducer system for voltage measurement up to + 5 kV, accurate to within 0.1%. Conceptually the instrument is identical to the field mill system shown in Figure 5.7 but has the advantage that the vibrating shutter system can be realised in a form which is physically very compact (Vosteen, 1974).

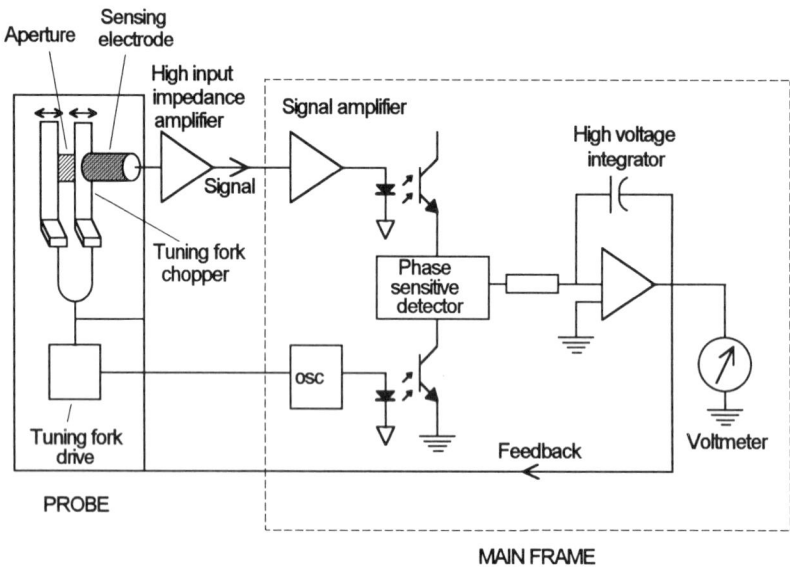

**Figure 5.8**    *Schematic diagram of Isoprobe electrostatic voltmeter (Monroe Electronics Inc.).*

## 5.4  VOLTAGE DIVIDERS

One of the commonest techniques used for measuring high voltage is to create an attenuated version of the voltage signal to be monitored by means of a divider network. For direct high voltages - the most usual output from the supplies connected to electrostatic systems - resistive dividers are employed. For alternating voltage measurement, resistive dividers may be specified, but for accuracy and lower power dissipation, capacitive dividers are in many ways more satisfactory. Where impulse/stepfunction potential signals are involved, careful analysis of the resistance, inductance and capacitance of the divider elements is required to assess how much distortion the divider is likely to introduce into the attenuated output signal.

Potential dividers have certain physical features in common, irrespective of the signal attenuating system. First of all, connection of the high voltage input is by way of a terminal designed to minimise the possibility of corona - usually the terminal is configured as a toroid, sphere or thick disc with rounded edges (see Figure 5.9). Secondly, the divider would normally be enclosed to shield it from environmental moisture and dust and to act as a physical barrier to any free, airborne charges from nearby ion sources, e.g. corona points. Clearly, attenuation ratio errors arise if a current passing through the surrounding space enters the divider system partway along its length. Finally, the low-voltage divider-output signal is fed to a display instrument. Normally a buffer amplifier is interposed between the divider output and the display device, so that there is no appreciable current demand on the divider which would result in erroneous measurements.

**Figure 5.9**    *Different types of high voltage terminals.*

### 5.4.1 Direct Voltage Dividers

While in concept resistive voltage dividers are extremely simple devices, in practice maintaining a consistent attenuation ratio over a range of operating temperatures is difficult to achieve with input voltages spanning perhaps a 30:1

ratio. The problem is that the requirement for the divider to draw a low current means that the divider elements require very high resistance. Such elements may not be strictly ohmic, i.e. their current/voltage characteristic may be somewhat non-linear.

In relatively low cost dividers, a significant number of standard 0.5 W high-stability carbon film resistors may be connected in series and mounted on an insulated support frame (often of polymethylmethacrylate) to form a helical spiral as shown in Figure 5.10. An insulating outer cover is provided for the resistor stack.

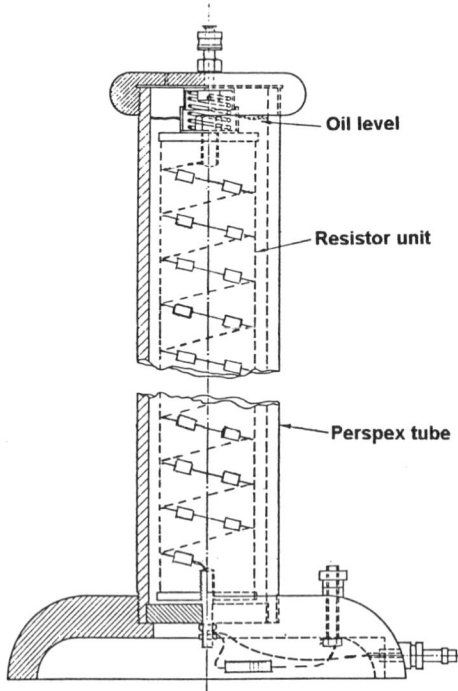

**Figure 5.10**    *Low cost resistor-divider unit*

Typically the resistor elements would be specified to give a maximum divider current of 10 μA, requiring a resistance/rated-voltage value of $10^5$ Ω/V. Individual resistors, with a notional voltage rating of 500 V, would then be operated at not more than 100 V, thus requiring resistance values of at least 10 MΩ in which the power dissipation would be 1 mW. To avoid internal corona action, the whole unit may be filled with transformer oil. Modern high-stability carbon film resistors of high resistance value have a temperature coefficient up to ± 1000 ppm/°C, and a resistance drift of not more than 2% per year. Dividers accurate to between 2% and 3% of maximum reading over a practical range of operating temperatures can be realised without too much trouble, but to obtain long-term, consistent performance within tighter limits is more demanding.

Metal film resistors fabricated from an evaporated nickel-chromium layer spirally grooved to increase the length of the current path, and thus the end-to-end resistance, have a better temperature coefficient ($\pm$ 50 ppm/°C) than high-stability carbon film units. But, with a maximum resistance in standard 0.25W resistors of 1 M$\Omega$, more units are required than with carbon film to construct a divider for any specific maximum voltage. Special 1 W metal film resistors, individually rated at up to 10 kV and with resistance values up to 100 M$\Omega$ (Welwyn Electric type M.40) have been developed for incorporation into high voltage measurement systems. Since these resistor components display only a small resistance variation from zero to full load, it is possible to achieve consistent divider performance accurate to within 1%.

Where divider accuracy better than 1 part in $10^3$ is required, wire-wound resistors, based on one of the nickel/copper alloys, should be specified. Individual non-inductively wound resistor elements of 1 M$\Omega$ maximum resistance are spirally mounted as previously described, and oil immersed. The excellent stability of such units ($\pm$3 ppm/°C, long term resistance variation < 30 ppm/year) is offset to some extent by their higher current demand, since it is too expensive and hence impractical to specify resistance/rated-voltage ratios exceeding about $10^4$ $\Omega$/V.

Bowdler (1973) recommends that, during construction, the divider ratio of multi-element divider units should be established by connecting the N individual resistors in parallel and measuring the resultant resistance $R_P$ accurately. If all the individual elements have resistance values within 1% of the mean, the resistance $R_S$ of the series-connected stack is given by

$$R_S = N^2 R_P \qquad (5.8)$$

accurate to better than 1 part in $10^4$. The resistance of the $N^{th}$ element, i.e. the resistor connected to ground in Figure 5.11, needs to be measured accurately (to within 1 part in $10^4$) to realise a divider with a measurement capability accurate to a few parts in $10^4$, given stable resistors and careful design and assembly. Where there is concern to obtain the ultimate divider accuracy, a temperature-stabilised enclosure for the resistor stack may be specified. A multiple guard electrode structure, coupled to a secondary divider, minimises the probability of local corona action or leakage to or from the primary divider stack (Figure 5.12).

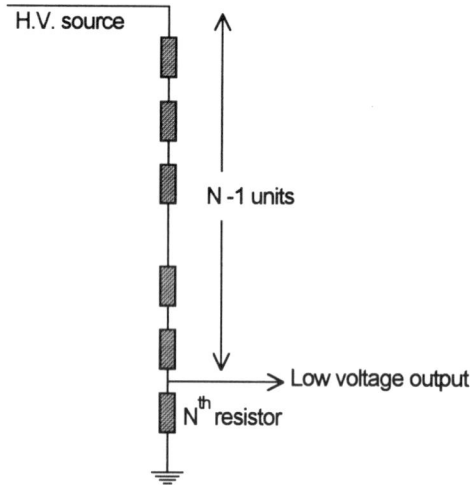

**Figure 5.11**    *A resistor divider.*

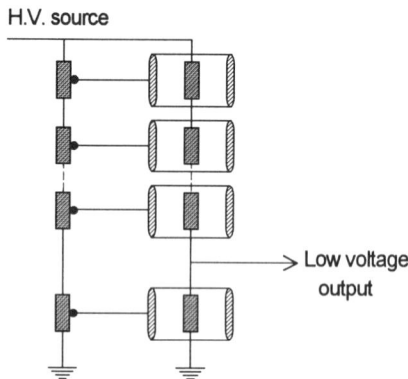

**Figure 5.12**    *Screening system coupled to a secondary divider.*

Resistive dividers with no matching capacitive elements are at risk when subjected to transient voltages (source energisation for example). The voltage division on the resistor stack immediately following the arrival of the voltage wavefront is determined by the stray capacitances of the divider structure. This may result in the upper resistor elements being overstressed since the lower elements have a greater stray capacitance to ground. This problem can at least be alleviated by fitting a physically large high-voltage terminal structure, thereby giving a more uniform capacitance grading down the stack.The voltage output from the divider during surge excitation of the high voltage terminal is a complex function of the resistance, inductance and capacitance of the stack structure. High transient voltages may be generated. It is important, therefore, to incorporate protection circuits on the buffer amplifier input (section 7.3.1.3).

### 5.4.2 Alternating Voltage Dividers

Since alternating sources are not used extensively with electrostatic equipment (exhaust gas precipitators excepted), dividers for such voltage measurement will only be treated briefly. For a detailed review of this topic see Bowdler (1973) and Kuffel and Abdullah (1970).

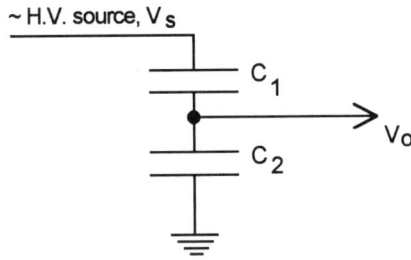

**Figure 5.13**  *A conceptual capacitor divider.*

While resistive dividers can be used at low frequency (50Hz) and up to about 100 kV, increasing frequency and voltage give rise to an unacceptable fall-off in accuracy, due to both the phase and amplitude errors at the divider output. These errors are associated with the distributed stray capacitance to earth from the stack, and the capacitive coupling between individual elements.

Clearly, alternating voltage dividers with low current demand can be realised by using a capacitor configuration such as is shown in Figure 5.13 where

$$V_O = V_S \frac{C_1}{C_1 + C_2}. \tag{5.9}$$

Ideally, $C_2$ is a capacitor of high stability and low loss, although it may well be shunted by a high value resistance to ensure that no residual direct voltage can be left at the divider output point. $C_1$ should also be a low loss capacitor, but of much smaller capacitance than $C_2$, so giving the required divider ratio. For a simple voltage monitoring system, $C_1$ may be merely the capacitance between two appropriately separated sphere electrodes, or a series connection of low voltage capacitor units. With such systems, repositioning of nearby earthed objects can influence the effective value of $C_1$, so modifying the divider ratio.

For more consistent measurements, but retaining the constraint of low current demand, special screened capacitor units may be constructed such as that shown in Figure 5.14. The central portion of the low-voltage electrode operates with uniform-field, corona-free conditions over its whole surface. The upper and lower parts of the low voltage electrode effectively decouple the well-defined measurement capacitor (centre portion of low voltage electrode/coaxial high voltage electrode) from the surrounding environment.

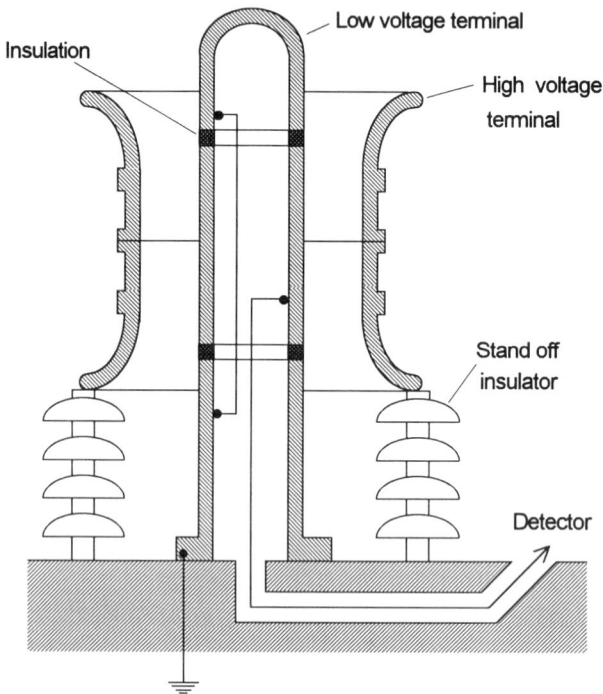

**Figure 5.14**    *High voltage capacitor with screening electrode.*

Screened capacitors are expensive mechanical assemblies which may be minimised in size by pressurising, and by careful choice of the radii ratio of their coaxial electrodes. For a coaxial capacitor (Figure 5.15) the maximum field, $E_M$, occurs at the surface of the inner electrode. Assuming $E_M$ is the highest field that can be maintained without corona or breakdown, the maximum sustainable voltage, $V_M$, is given by the expression

$$V_M = E_M r_1 \ln\left(\frac{r_2}{r_1}\right). \qquad (5.10)$$

For a given capacitor size ($r_2$ fixed) it is easily shown that choosing $r_1 = r_2/e$ (where $e = 2.718$) gives the highest value for $V_M$, i.e.

$$V_M = \frac{E_M r_2}{2.718}. \qquad (5.11)$$

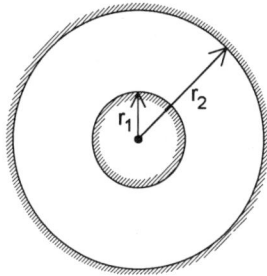

**Figure 5.15**   *Plan view of a coaxial high voltage capacitor.*

### 5.4.3  Dividers for Impulse and Step-Function Voltages

The dividers discussed so far have been assumed to be connected across either invariant direct voltage or steady-state alternating voltage supplies. In practice, rapid change of voltage (switch-on/switch-off) or the requirement to monitor accurately both the peak amplitude and time variation of a voltage pulse place special constraints on dividers. Essentially, it is necessary to devise a signal attenuation system which maintains an invariant divider ratio over an extended frequency range right from zero (direct voltage).

Direct voltage resistive dividers may have capacitors added to ensure that stray capacitance has only a second order effect on the divider output, so providing accurate voltage monitoring during both supply energisation and subsequent variation. For a mixed resistance-capacitance divider (Figure 5.16), the required conditions are

(i) $C_1 \gg$ stray capacitance     (ii) $C_2/C_1 = R_1/R_2$ .

In general, a more detailed analysis of the system components including the divider itself, the connection to the high voltage source, and the output cable to the monitoring device (typically an oscilloscope) is necessary in order to assess how accurately the source signal is recorded. Specifically, the effect of distributed inductance needs to be taken into account, since this can give rise to short-term variations in the effective divider ratio, or can interact with the divider capacitance to initiate high frequency resonance. The divider response (i.e. low voltage output signal) for various assumed conditions is illustrated in Figure 5.17.

A formal theoretical treatment of the possible divider response is set out in the papers by Bowdler (1973) and Kuffel and Abdullah (1970).

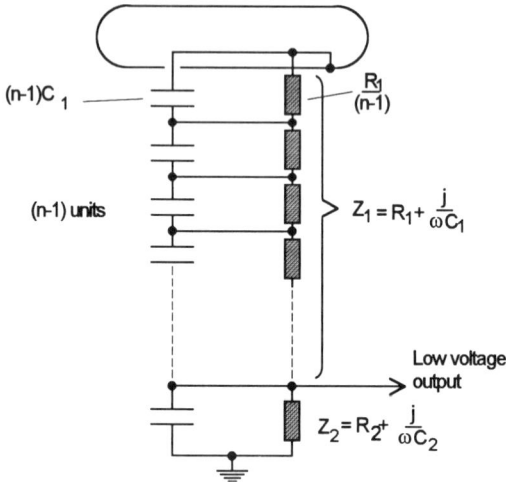

**Figure 5.16**    *A composite resistor/capacitor divider.*

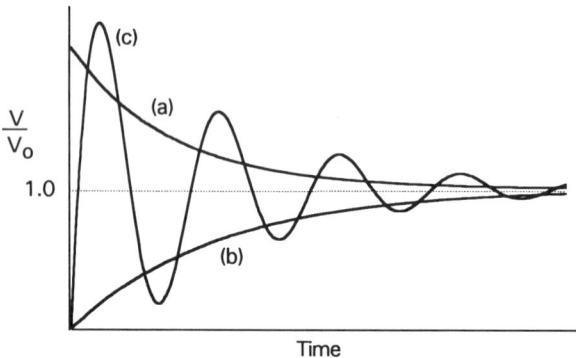

**Figure 5.17**    *Divider response to application of step-function voltage. The notional output voltage is $V_0$. Resistor divider with (a) inductive low-voltage impedance $Z_2$, (b) inductive high-voltage impedance $Z_1$ and (c) a capacitive divider with series inductance in high-voltage impedance.*

## 5.5 SPARK GAPS

In high voltage laboratories, breakdown between electrodes of well-defined shape and location has long provided a useful and relatively simple method of peak voltage measurement. Conventionally, spherical electrodes have been employed, and detailed tables are available showing the relationship between sphere diameter, D, inter-electrode gap s, and sparkover voltage (BS 358:1960). Part of such tabulated data is shown in Table 5.1. For alternating voltages, measurements accurate to $\pm3\%$ are possible if s<0.5D while for direct voltages, accuracy to within $\pm5\%$ is claimed for s<0.4D.

A number of factors are found to influence the breakdown voltage, causing deviations from the value given for a particular electrode size and inter-electrode gap. If earthed objects are located within the zone of influence of the breakdown gap, a progressive reduction of sparkover voltage is observed as the earthed object is brought closer to the gap. The probability of gap breakdown is obviously linked with both the likelihood of a breakdown-initiating electron appearing in the high field zone between the electrodes, and to a subsequent "ionisation-runaway" condition developing. Provision of an initiating electron can be facilitated by irradiating the inter-electrode gap, so tending to reduce the breakdown voltage. Conversely, addition of water vapour (increased humidity) increases the breakdown voltage (Kuffel, 1961). This latter effect is associated with ionisation suppression due to enhanced electron attachment. Dust and draughts need to be avoided if further variations of the breakdown voltage are not to be in evidence.

Tabulated values of sphere-gap sparkover voltages (such as those in Table 5.1) are quoted for standard pressure and temperature conditions (1013 millibars and 20°C) and due allowance must be made for different ambient conditions. For a pressure p mbars and temperature T °C, an air density factor $\rho$ can be defined by

$$\rho = 0.29\frac{p}{273 + T}. \qquad (5.12)$$

Figure 5.18 shows the relationship between the breakdown voltage adjustment factor, A, and the air density factor $\rho$. The effective breakdown voltage $V_e$ is then related to the breakdown voltage,under standard pressure and temperature conditions, $V_s$ by the relation

$$V_e = AV_s. \qquad (5.13)$$

**Table 5.1**   *Flashover voltages for alternating and direct voltages between spherical electrodes.*

| Sphere-gap spacing (cm) | Kilovolts peak at 20°C and 1013 millibars | | | | | | | | | | | |
|---|---|---|---|---|---|---|---|---|---|---|---|---|
| | Sphere diameter (cm) | | | | | | | | | | | |
| | 2 | 5 | 6.25 | 10 | 12.5 | 15 | 25 | 50 | 75 | 100 | 150 | 200 |
| 0.05 | 2.8 | | | | | | | | | | | |
| 0.10 | 4.7 | | | | | | | | | | | |
| 0.15 | 6.4 | | | | | | | | | | | |
| 0.20 | 8.0 | 8.0 | | | | | | | | | | |
| 0.25 | 9.6 | 9.6 | | | | | | | | | | |
| 0.30 | 11.6 | 11.2 | | | | | | | | | | |
| 0.40 | 14.4 | 14.3 | 14.2 | | | | | | | | | |
| 0.50 | 17.4 | 17.4 | 17.2 | 16.8 | 16.8 | | | | | | | |
| 0.60 | 20.4 | 20.4 | 20.2 | 19.9 | 19.9 | | | | | | | |
| 0.70 | 25.8 | 23.4 | 23.2 | 23.0 | 23.0 | | | | | | | |
| 0.80 | 28.3 | 26.3 | 26.2 | 26.0 | 26.0 | | | | | | | |
| 0.90 | 30.7 | 29.2 | 29.1 | 28.9 | 28.9 | | | | | | | |
| 1.0 | (35.1) | 32.0 | 31.9 | 31.7 | 31.7 | | | | | | | |
| 1.2 | (38.5) | 37.6 | 37.5 | 37.4 | 37.4 | | | | | | | |
| 1.4 | (40.0) | 42.9 | 42.9 | 42.9 | 42.9 | | | | | | | |
| 1.5 | | 45.5 | 45.5 | 45.5 | 45.5 | 45.5 | | | | | | |
| 1.6 | | 48.1 | 48.5 | 48.1 | 48.1 | 48.1 | | | | | | |
| 1.8 | | 53.0 | 53.5 | 53.5 | 53.5 | 53.5 | | | | | | |
| 2.0 | | 57.5 | 58.5 | 59.0 | 59.0 | 59.0 | | | | | | |
| 2.2 | | 61.5 | 63.0 | 64.5 | 64.5 | 64.5 | | | | | | |
| 2.4 | | 65.5 | 67.5 | 69.5 | 70.0 | 70.0 | 70.0 | | | | | |
| 2.6 | | (69.0) | 72.0 | 74.5 | 75.0 | 75.5 | 75.5 | | | | | |
| 2.8 | | (72.5) | 76.0 | 79.5 | 79.5 | 80.0 | 81.0 | | | | | |
| 3.0 | | (75.5) | 79.5 | 84.0 | 85.0 | 85.5 | 86.9 | | | | | |
| 3.5 | | (82.5) | (87.5) | 95.0 | 97.0 | 98.0 | 99.0 | | | | | |
| 4.0 | | (88.5) | (95.0) | 105 | 108 | 110 | 112 | 112 | 112 | | | |
| 4.5 | | | (101) | 115 | 119 | 122 | 125 | 125 | 125 | | | |

**Table 5.1** (cont.)

| | | | | | | | | | |
|---|---|---|---|---|---|---|---|---|---|
| 5.0 | (107) | 123 | 129 | 133 | 137 | 138 | 138 | | |
| 5.5 | | (131) | 138 | 143 | 149 | 151 | 151 | | |
| 6.0 | | (138) | 146 | 152 | 161 | 164 | 164 | 164 | 164 |
| 6.5 | | (144) | (154) | 161 | 173 | 177 | 177 | 177 | 177 |
| 7.0 | | (150) | (161) | 169 | 184 | 189 | 190 | 190 | 190 |
| 7.5 | | (155) | (168) | 177 | 195 | 202 | 203 | 203 | 203 |
| 8.0 | | | (174) | (185) | 206 | 214 | 215 | 215 | 215 |
| 9.0 | | | (185) | (198) | 226 | 239 | 240 | 241 | 241 |
| 10.0 | | | (195) | (209) | 244 | 263 | 265 | 266 | 266 | 266 |

Advocates of the use of uniform field electrode systems for measuring voltage by means of sparkover point out the inherent problem with sphere gaps - that the breakdown path is restricted to a small region about the common electrode axis, which is the zone where the highest field exists. By contrast, in uniform field systems, i.e. electrodes with Bruce or Rogowski profiles, the high field extends over a significant proportion of the electrode surface, thereby increasing the probability of finding an initiating electron for the breakdown process, and so reducing the scatter in breakdown voltage values recorded. Thus,after some initial conditioning sparkovers, scatter in breakdown voltage values is found to be within ±0.2% of the mean value, provided dust particles are not allowed to fall on the

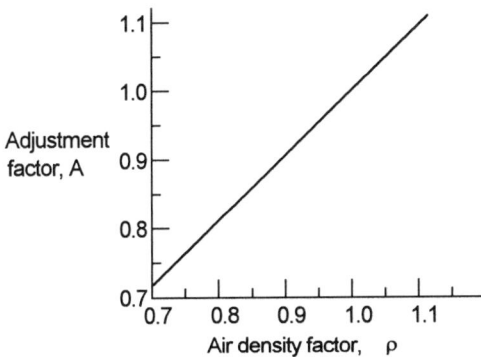

**Figure 5.18** *Voltage breakdown adjustment factor to allow for pressure and temperature changes.*

electrode surfaces (Kuffel, 1961). Qualitatively similar variations in breakdown values as with sphere gaps are observed for humidity changes, but gap irradiation has less effect as expected.

In air at pressure p mbar and temperature T °C, the breakdown voltage of a uniform field gap is given by the expression (Boyd et al, 1966)

$$V_S = 24.49\rho s + 6.61\sqrt{\rho s} \qquad (5.14)$$

where s is the gap spacing in cm and $\rho$ is the relative air density as defined by equation (5.12). To minimise scatter, s must be held below 0.8F, where F is the diameter of the flat portion of the uniform field electrode.

While sparkover between spheres or uniform field electrodes can provide a fairly accurate measure of voltage under laboratory conditions, its use for field-service measurements on electrostatic equipment operating up to about 100 kV provides less satisfactory and less accurate performance than direct-reading instruments. In practice, the inter-electrode gap is set wider than that corresponding to breakdown for the expected operating voltage. The spacing is reduced in increments until breakdown occurs repeatedly. Voltage measurement accuracy using such methods should be within 10% of the true value (Bright et al, 1978).

## 5.6 OTHER TECHNIQUES FOR VOLTAGE MEASUREMENT

In many practical situations, it is important to obtain a good indication of conductor potential without making a direct connection to the measuring device. Providing the measurement system has a very high input impedance, effective "coupling" to the conductor can be attained by

(a) placing a weak radioactive source in the probe to induce air-ionisation in the probe-conductor interspace.

(b) fixing a sharp point, e.g. a pin or a needle, to the probe such that corona discharging can occur at its tip. Typically the conductor/tip potential difference is less than 4kV.

(c) allowing water to drip from the probe. Charges are induced in and carried away by each water drop until the probe is in a zero-field situation,i.e. at the same potential as the nearby conductor.

Since the charging current is low for these techniques, especially in case (c), the time to reach steady-state can be quite extended. It is important to ensure that no significant charge leakage occurs in the measurement system. This requires the overall probe/coupling-lead insulation to be of very high integrity.

A variant on the non-contact measurement systems discussed above is to connect the probe to a variable high voltage source via a sensitive current meter. If the source output voltage is varied so as to plot a current/voltage characteristic about the zero-current point, the conductor potential can be inferred. Clearly the current meter needs to have built-in overload protection (section 7.3.1.3), while the use of a radioactive-source probe would be inadvisable when there is a possibility of energetic conductor-to-probe sparkover. Figure 5.19 shows a typical current/source voltage plot for a corona-point measurement system. There are conceptual similarities between this measurement method based on a source voltage, and the field mill/isoprobe systems discussed in section 5.3, in that both attain a balance state when there is zero field at the sensor/probe surface.

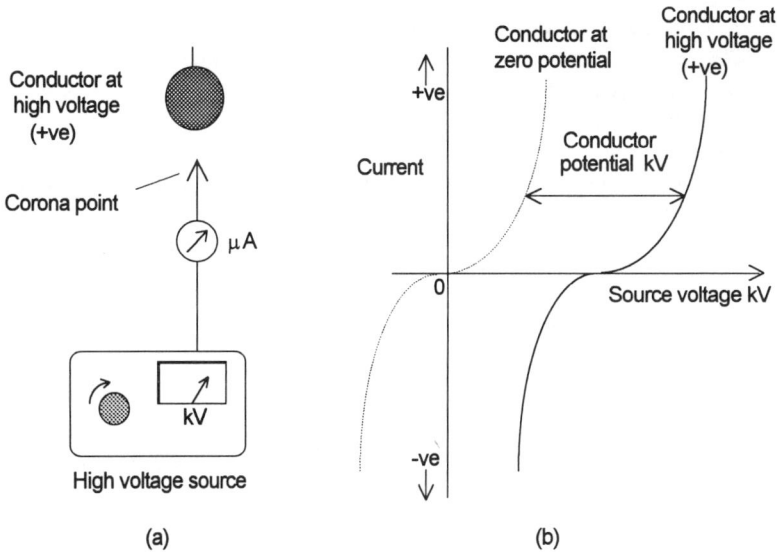

**Figure 5.19** *Using a corona source for voltage measurement. (a) The experimental arrangement and (b) the shift in the corona current-voltage characteristic when the conductor voltage is raised to high positive potential..*

Van der Weerd (1971) has demonstrated an extremely useful technique for plotting space potential, using a field meter sensing head as a potential probe. The distortion of the local space potential resulting from the introduction of the earthed field meter sets up a field at the sensing head which is found to be a function of the undisturbed potential with no field meter present. The local space potential, $V_\rho$, is then related to the measured electric field $E_m$, and the diameter D of the sensing head, by the equation

$$V_\rho = f.(E_m D) \qquad\qquad (5.15)$$

where the adjustment factor $f$ is approximately unity. The relationship breaks down if the fieldmeter sensing head is brought close (i.e. d<5D) to any conducting surfaces.

**REFERENCES**

Bowdler GW 1955 "An attracted-disc absolute voltmeter" *Proc IEE Part B* 301-312.

Bowdler GW 1973 *"Measurements in high-voltage test circuits"* (Oxford: Pergamon Press).

Boyd HA, Bruce FM and Tedford DJ 1966 "Sparkover in long uniform-field gaps" *Nature* **210** 719-720.

Böcker H 1939 "Voltmeter for 600 kV" *Arch f Elecktrotech* **33** 801- 810.

Bradshaw E, Husain SA, Kesavamurty N and Menon KB 1948 "Absolute measurement of high voltages by oscillating electrode systems" *JIEE* **95 Part II** 636-644.

Bright AW, Corbett RP and Hughes JF 1978 *"Engineering Design Guide 30:Electrostatics"* Oxford University Press.

Brooks HB, Defandorf FM and Silsbee FB 1938 "Absolute electrometer for the measurement of high alternating voltages" *J Res Nat Bur Stand* **20** 253-316.

Bruce FM 1947a "Calibration of uniform-field spark-gaps for high voltage measurement at power frequencies" *Proc IEE* **94 Part II** 138-149.

Bruce FM 1947b "The design of an ellipsoid voltmeter for precision measurement of high alternating voltages" *Proc IEE* **94 Part II** 129-137.

Kelvin (Lord) 1884 *"Papers on electrostatics and electromagnetism"* London.

Kuffel E 1961 "Influence of humidity on the breakdown voltage of sphere-gaps and uniform-field gaps" *Proc IEE* **108A** 295-301; 314-316.

Kuffel E and Abdullah M 1970 *"High-voltage engineering"* (Oxford: Pergamon Press).

Secker PE 1975 "The use of field mill instruments for charge density and voltage measurement" *Inst Phys Conf Ser No 27* 173-181.

Trump JG, Safford FJ and Van de Graaff RJ 1940 "Generating voltmeter for pressure-insulated hv sources" *Rev Sci Instrum* **11** 54-56.

Van der Weerd JM 1971 "Electrostatic charge generation during washing of tanks with water sprays : II Measurements and interpretation" *Inst of Phys Conf Ser No11* 158-177.

Vosteen RE 1974 "D.C.electrostatic voltmeters and field meters" *Conf Record 9th Annual Meeting IEEE/IAS* 799-810.

Waterton FW 1976 "A 300kV electrostatic voltmeter" *J Phys E: Scientific Instruments* **9** 647-650.

# CHAPTER SIX

# MEASUREMENT OF CHARGE

As we saw in Chapter 3 the majority of electrostatic processes require the controlled charging of particles and/or surfaces, and the subsequent exploitation of the resulting electrostatic force effects. An ability to measure charge, whether on particles, surfaces of sheet products or distributed throughout a volume as an ion cloud or charged mist, is extremely important therefore in understanding and optimising the operative mechanisms, and quantifying the effectiveness of the electrostatic processes.

## 6.1 CHARGED PARTICLES

From Chapter 2 we know that when a charged particle experiences an electric field, E, the electrostatic force, F, acting on it is given by

$$F = qE \tag{6.1}$$

where q is the charge on the particle. The charge is generally maintained on the surface of the particle and may well be distributed uniformly over it. The maximum charge which can be maintained on the particle is controlled by the breakdown strength, $E_{BD}$, of the surrounding environment. If the particle is spherical, has radius a and charge q, the surface field in an environment of relative permittivity $\varepsilon$ is

$$E_{SURF} = \frac{q}{4\pi\varepsilon\varepsilon_0 a^2}. \tag{6.2}$$

Putting $E_{SURF}$ equal to $E_{BD}$ then gives the maximum charge on the particle, i.e.

$$q_{MAX} = 4\pi\varepsilon\varepsilon_0 a^2 E_{BD}. \tag{6.3}$$

Where electric fields are to be used to direct the movement of charged particles, it is essential that electrostatic rather than gravitational forces should predominate. This condition can be described by an effectiveness ratio R, where

$$R = \frac{\text{Electrostatic Force}}{\text{Gravitational Force}} = \frac{qE}{mg}. \qquad (6.4)$$

For a given applied field E and with q at its maximum value, $q_{MAX}$,

$$R_{MAX} = \frac{4\pi\varepsilon\varepsilon_0 a^2 E_{SURF}}{(4\pi a^3/3)\rho g} E = \frac{3\varepsilon\varepsilon_0 E_{SURF}}{a\rho g} E \qquad (6.5)$$

where $\rho$ is the particle density. Clearly the effectiveness ratio increases as the particle radius, a, is reduced, showing why particle-associated electrostatic processes make use of solid matter and liquid drops in as finely divided form as possible.

From equation (6.4) we see that the important parameter in determining the effectiveness ratio is (q/m) the charge-to-mass ratio of the particles rather than the mean charge alone. Traditionally, the charge is measured in a Faraday pail. Thereafter, the particles may be weighed and (q/m) determined.

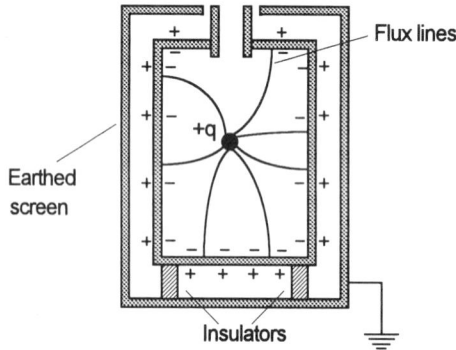

**Figure 6.1**    *Cross-section through a cylindrical Faraday pail.*

A cross-sectional view of a typical Faraday pail is given in Figure 6.1. It consists of a closed conducting chamber with a small entrance orifice through which particles can enter. The pail is isolated from ground and normally has an earthed screening vessel surrounding it to prevent external charges from being measured and to reduce noise from extraneous electrical sources. If a particle with charge +q enters the Faraday pail it induces a charge -q distributed over the inner surface of the isolated chamber. To retain the electrical neutrality of this isolated chamber, a positive charge +q is distributed over its outer surface, setting up an electric field and, therefore, a potential difference between it and the earthed outer screen. If the capacitance between the Faraday pail and its screen is $C_P$ then the pail acquires a potential

$$V_F = \frac{q}{C_P} \tag{6.6}$$

which can be measured by any high input-impedance voltmeter. It is usual, however, to measure the charge directly with a virtual earth amplifier (section 4.2) thereby holding the pail essentially at earth potential. In this situation (see Figure 6.2) the charge +q on the outer surface of the pail flows to the feedback capacitor $C_f$ which results in the amplifier output changing by $-\Delta v_0$, where

$$\Delta v_0 = \frac{q}{C_f} \tag{6.7}$$

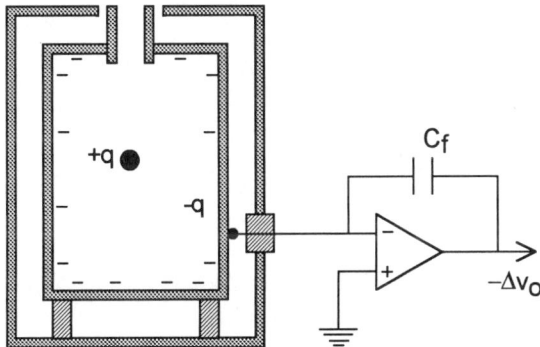

**Figure 6.2**    *Faraday pail with virtual earth feedback amplifier.*

We should note that when measuring charges using a Faraday pail it is not necessary for the particle to contact the walls of the pail or for actual charge transfer to occur from the particle to the wall. Consequently, the technique works equally well for both conducting and insulating particles. The measurement is based on an induced charge effect; for high accuracy this requires that the field lines from the particle terminate only on the inner walls of the pail so that

$$|q_{induced}| = |q_{particle}|.$$

Where a group of particles can be identified as the only charge species present, and the particles follow a predetermined trajectory, it is possible to use a cylindrical induced-charge detector (Corbett and Bassett, 1971) of the form shown in Figure 6.3.

If each particle passing through the detector has a charge of $10^{-13}$ C and the feedback capacitor has a value of 500 pF, individual particles passing through the detector would result in pulses of magnitude 0.2 mV which may be observed on the oscilloscope. It is important to note that, in this cylindrical detector, contact between the charged particle and the cylinder surface must be avoided otherwise erroneous results may occur due to contact charging.

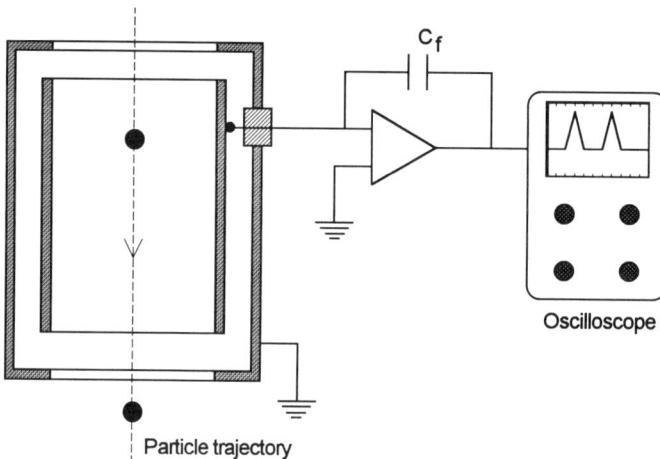

**Figure 6.3**    *An induced-charge meter for particles based on the Faraday pail.*

One of the principal difficulties with measuring particle charge is that the particles of interest are often accompanied by a host of free ions created during the particle charging process. For example, in a powder coating gun which uses a high-voltage corona charging source, up to 99% of the charge released from the corona may remain as free ions. Under such circumstances the direct use of a Faraday pail leads to erroneous data since very many free ions enter the pail along with the charged particles. In consequence, the (q/m) ratio is greatly overestimated and may lead to problems where the behaviour of a coating system is being assessed in terms of its charging efficiency.

In principle, the problem of free-ion error can be overcome by placing an earthed, coarse-mesh grid over the aperture to the Faraday pail as in Figure 6.4. The free ions, having a high charge-to-mass ratio, will exhibit trajectories that closely follow the field lines and thus tend to flow to the mesh structure. By contrast, most of the charged powder particles, which have a greater mass and momentum and thus a lower charge-to-mass ratio, pass through the mesh into the Faraday pail.

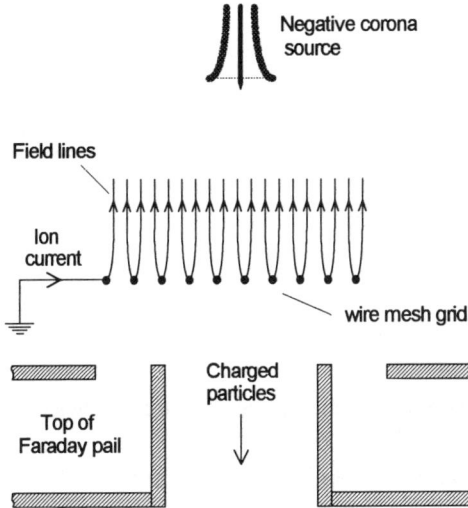

**Figure 6.4**   *Addition of an earthed wire mesh grid at the mouth of a Faraday pail acts as a block to free ions which are attracted to the mesh.*

The earthed-mesh modification may only produce a temporary alleviation of ion-associated errors. Gradually, a particle deposit builds up on the mesh surface and blocks the flow of ions to the mesh. The space charge field of the charged particles and ions approaching the Faraday pail is significantly enhanced in the neighbourhood of the mesh wires. Free ions collecting on the insulating powder layer coating the mesh may well cause back-ionisation to occur, with the ions so released changing drastically the charge on the in-bound particles.

A permanent solution is to arrange the mesh grid in a vertical plane as in Figure 6.5 and to allow water to trickle continuously over it, carrying away any deposited particles (Moyle and Hughes, 1982). With this system the charge measured by the electrometer circuit can be associated reliably with the particles collected on the L-shaped electrode. Thus, after weighing the collected particles, a realistic and consistent value of charge-to-mass ratio can be calculated. This system is, however, very significantly removed from the simple, portable Faraday pail that is initially assumed sufficient for the measurement of particle charge.

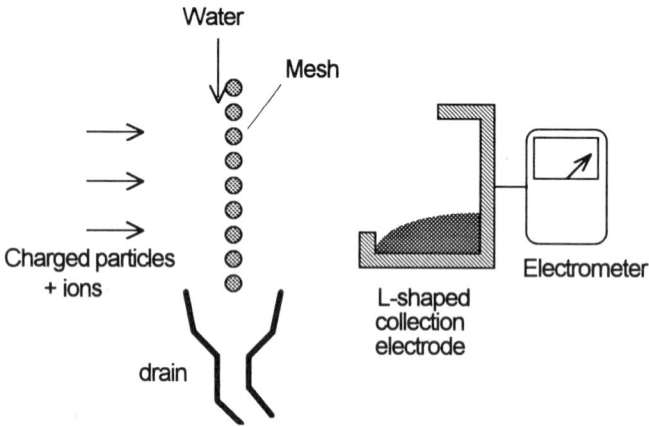

**Figure 6.5**    *A water-irrigated mesh grid to prevent particle accumulation on the grid.*

Other problems arise when attempting to monitor charge associated with a continuously flowing product. Suppose, in this case, material flows into the Faraday pail at some constant rate M kg/s. Assuming a constant charge-to-mass ratio on the product, the output voltage of the virtual earth amplifier will increase with time, t, since $V_F$ is now given by

$$V_F = \frac{q}{m}\left(\frac{Mt}{C_P}\right).$$    (6.8)

Clearly, after some time the output voltage from the amplifier (Figure 6.2) will saturate unless the measurement is discontinued periodically while the capacitor is shortcircuited. An alternative approach is to replace the capacitor by a feedback resistor and measure the current $I_p$ (section 7.3.1.3) flowing to ground through the pail. Now if the mass flow rate, $\dot{M}$, is known, (q/m) may be determined easily since

$$\frac{q}{m} = \frac{I_P}{\dot{M}}.$$    (6.9)

Of course, the measurement is still subject to the free-ion errors discussed above. A number of commercially-available electrometers are furnished with switchable ranges that enable a Faraday pail to be used either in the charge or current mode.

There are situations, for example the bulk transport of powder, flake or liquid from one container to another, where it is not practical to use a traditional Faraday pail because of the enormous volume of material being handled. However, it may still be possible to measure the charge on the material. As an example, consider the common industrial situation depicted in Figure 6.6(a) in which solid particulates are being pneumatically transferred from a silo into a shipping container. The walls of the shipping container are metallic and therefore can form the collecting chamber of a very large Faraday pail.

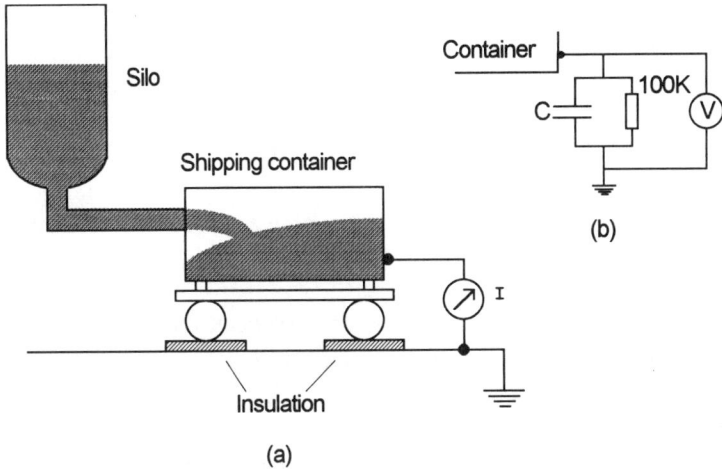

(a)

(b)

**Figure 6.6**     *The charging of particulates during pneumatic transfer from a silo to a shipping container may be measured as shown in (a). The rate of charge collection can be measured with a simple circuit in (b) where C is chosen to filter out mains frequency.*

It would be normal practice for the container to be securely earthed during the loading operation using a stout copper cable. If, instead, the wheels of the trailer supporting the container are placed on an insulating plastic sheet and the trailer connected to earth through a current measuring device then a simple, unscreened Faraday pail has been formed. Since the container is effectively earthed, filling may proceed safely and,by measuring the current flow to earth during loading and combining this with the known mass flow rate, a value for (q/m) may be deduced.

In such a test, the current may be measured using a virtual-earth current detector as described above. Alternatively, a high input-impedance voltmeter (DVM) may be used to measure the voltage developed across a high value resistor, say 100kΩ, connected between the container and earth as shown in Figure 6.6(b). The value of the parallel capacitor is chosen to reduce the mains noise picked up by such a large unscreened structure, i.e. C must satisfy the condition $10^5 \times C \gg 20$ ms.

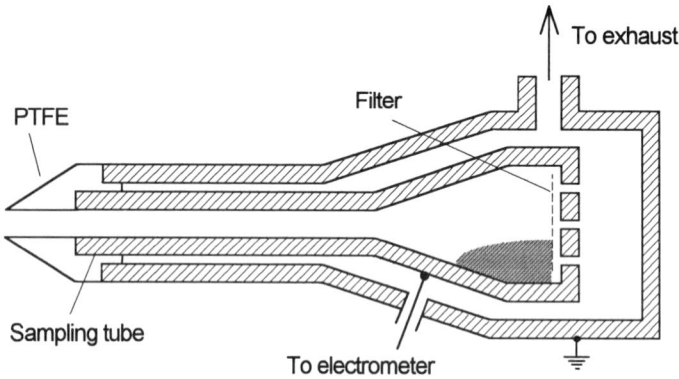

**Figure 6.7**    *Outline of the charge probe for measuring charge on powders extracted from a pneumatic conveyancing system.*

Variations of the Faraday pail concept have been developed to measure the charge on powder particles being conveyed pneumatically in ducts and there is particular interest in measuring (q/m) for dust particles in an electrostatic precipitator. Cross (1987) suggests the system shown diagrammatically in Figure 6.7. Charged particles are drawn through a sampling tube into the Faraday pail which contains a filter bag to separate the particles from the gas flow for subsequent weighing. The sampling tube must be conducting and must form part of the Faraday pail so that any contact charging resulting from the impact of ingested particles with the tube do not affect the measurement. Since the portion of the sampling tube projecting into the duct forms part of the Faraday pail, it is essential that it is screened. Otherwise, contact electrification resulting from particle impact on the outside of the sampling tube would give rise to errors. Finally, the inlet of the sampling tube requires careful design, so that sampling takes place under isokinetic conditions. In Cross' design, the correct aerodynamic shape of the inlet is achieved with a PTFE insert which also serves as an insulating spacer between the inner sampling tube and outer screen.

There must be some question concerning the use of such a system in a high speed gas flow where the outer surface of the PTFE insert could become charged by particle collisions. Such charge would be capacitively coupled to the sampling tube and would be measured by the Faraday pail. Additionally, the electrostatic field of the charged PTFE, by attracting particles of opposite polarity and repelling similarly charged particles, would alter the particle collection dynamics.

## 6.2 CHARGED SURFACES

In general, the measurement of surface charge density is effected with a field meter brought close to the charged surface. The intention is that the field meter should be the only earthed object coupling to the surface charge. When charges reside on both surfaces as in Figure 6.8, the field meter reading, E, is related to the *net* surface charge density $(Q_1-Q_2)/A$ by the relation

$$E = \left(\frac{1}{\varepsilon_0 A}\right)(Q_1 - Q_2) \tag{6.10}$$

Assuming that the material supporting the charge is a good insulator - perhaps a web of paper, plastic or rubber - it is likely that the high field within the material (due to opposing surface charges) will be of little practical consequence, giving rise neither to significant conduction nor to electrical breakdown. External to the material, the electric field associated with the charge can give rise to a range of phenomena (section 4.1).

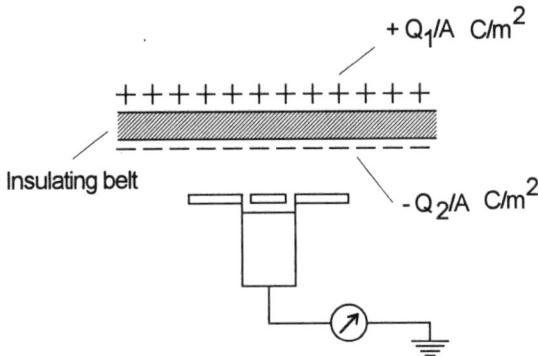

**Figure 6.8**    *Field meter measurement of the net surface charge on an isolated belt or web.*

In most cases, it is the net surface charge density or the external field (equation (6.10)) that is the parameter of interest. It is possible, however, to measure the surface charge density on the individual surfaces by reducing to negligible magnitude the external field associated with the charge on one of the surfaces. This may be achieved,as shown in Figure 6.9, by passing the web over an earthed, large diameter, conducting roller. The charge on the web surface in contact with the roller couples perfectly with induced charges on the roller itself and will not be "seen" by the field meter.

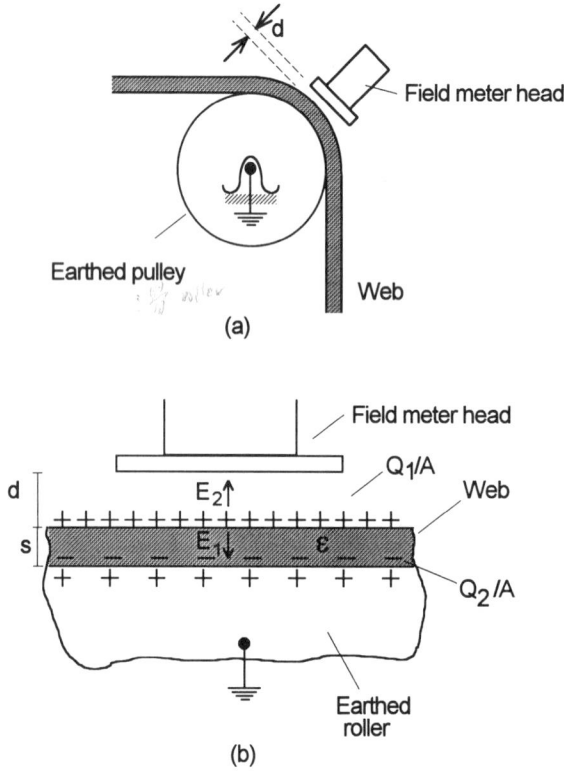

**Figure 6.9**    *Measurement of surface charge density on an individual web surface. (a) The physical arrangement and (b) a simplified model for analysis.*

The charge on the web surface remote from the roller couples to both the pulley and the earthed field meter. From equation (2.27) and noting that $E_1$ and $E_2$ are of opposite sign, we may write

$$\frac{Q_1}{A} = \varepsilon\varepsilon_0 E_1 + \varepsilon_0 E_2 \qquad (6.11)$$

and

$$E_2 d = E_1 s \qquad (6.12)$$

where s is the thickness of the web and d the distance from the web surface to the field mill. Combining equations (6.11) and (6.12) gives an expression for the required surface charge density, in terms of the measured field $E_2$, namely

$$\frac{Q_1}{A} = \varepsilon_0 E_2\left(\frac{\varepsilon d}{s} + 1\right). \tag{6.13}$$

For a surface charge density, $Q_1/A$, equal to $10^{-7}$ C/m² on a web of relative permittivity 2 and thickness 1 mm, the field $E_2$ varies with spacing d as shown in Figure 6.10.

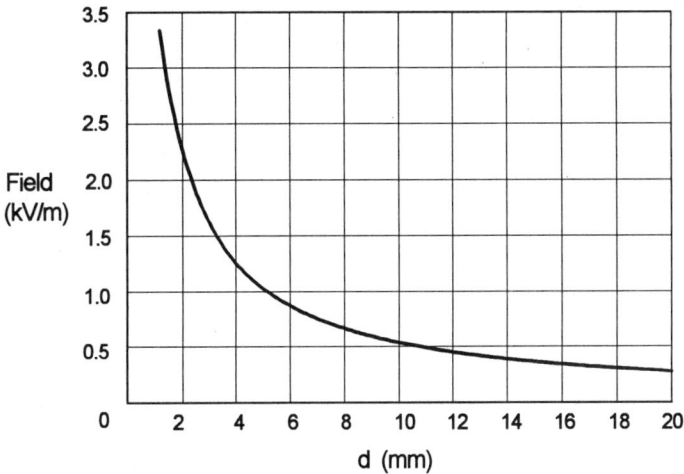

**Figure 6.10**  *Measured field at distance d from a web running over a large diameter earthed roller ($Q_1/A = 10^{-7}$ C/m², s = 1 mm, $\varepsilon = 2$).*

It is clear that $E_2$ is significantly attenuated when d>>s. Making d small enhances $E_2$ but also increases the percentage error in the estimated value for d. Thus field mills are not particularly suitable for this type of measurement; the problem is made worse by their ill-defined "field-sensing plane" (section 4.3.2) which limits the closest distance of approach if reasonable accuracy is to be maintained. A better approach is to use an induction probe (section 4.2) of relatively small diameter.

It is sometimes suggested that after measuring $Q_1/A$ the charge density $-Q_2/A$ may be evaluated by passing the web over a second roller which now contacts the surface on which $Q_1/A$ resides and measuring the field as before. While this approach is fine in principle it should be remembered that,when measuring $Q_1/A$, field-aided contact charging of the web by the roller may well have modified the value of -$Q_2$/A!  A better method is to measure the net surface charge density prior to measuring $Q_1$/A and then to use equation (6.10) to deduce -$Q_2$/A once $Q_1$/A is known.

### 6.2.1 Achieving Good Spatial Resolution in Surface Charge Measurements

While for many industrial situations it is only important to know the mean net surface charge density or resultant field, there are instances where a knowledge of the spatial distribution of charge is essential - for example, in studying charge migration over surfaces due to charge-density gradients, or looking at the charge pattern on an "exposed" photoconductor of an electrophotocopying machine. In such cases, a high resolution induction probe can be exploited, in which the sensing element is a fine wire surrounded by a guard electrode held at an appropriate potential.

Consider the situation where a thin insulating film is mounted directly on a grounded conductor, and its exposed face has a positive, position-variable charge density $\sigma_S$ . The probe/guard combination is mounted close to the surface and a high input-impedance unity-gain amplifier is used to drive the guard electrode to the same potential as the probe wire (see Figure 6.11(a)). To simplify analysis it is assumed that :

(i) The amplifier is ideal, i.e. no bias current and an infinite input impedance.

(ii) The probe is close enough to the sample surface and the guard electrode is so shaped that it is realistic to consider the field in the interspace between probe and sample surface to be uniform.

We can represent the probe/film interaction by the equivalent circuit shown in Figure 6.11(b). Assume that the area of the film seen by the probe is A. The surface charge on the sample ($q = q_1+q_2$), which gives rise to a potential $V_S$ on the sample surface, is shared between the sample capacitance, $C_S$, and the series combination of the air gap capacitance, $C_{ps}$, and the total capacitance to ground, $C_{pg}$,at the amplifier input. Now

$$q_2 = (V_S - V)C_{ps} = VC_{pg} \qquad (6.14)$$

which upon rearranging yields

$$V_S = \frac{C_{ps} + C_{pg}}{C_{ps}}.V \qquad (6.15)$$

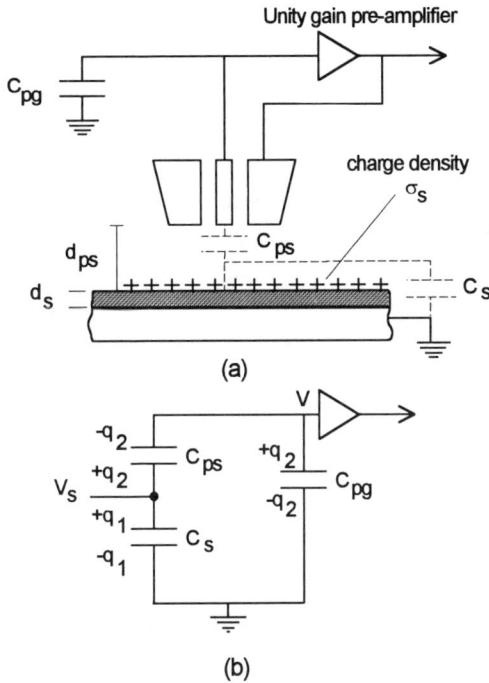

**Figure 6.11** *Scanning induction probe showing (a) the experimental arrangement and (b) a simple equivalent circuit.*

where V is the voltage appearing at the amplifier input. The surface charge density, $\sigma_S$, on the sample is given by

$$A\sigma_S = q_1 + q_2 = C_S V_S + C_{pg} V \qquad (6.16)$$

Substituting from equation (6.15) then gives

$$A\sigma_S = \frac{C_S(C_{ps} + C_{pg})}{C_{ps}} \cdot V + C_{pg} V \qquad (6.17)$$

which upon rearranging gives V directly, i.e.

$$V = \frac{A\sigma_S C_{ps}}{C_{ps}(C_S + C_{pg}) + C_S C_{pg}} \qquad (6.18)$$

In most practical cases $C_{ps} \ll C_{pg}$ and $C_S$. With this caveat, equation (6.18) simplifies to

$$V = \frac{AC_{ps}}{C_S C_{pg}}\sigma_S . \qquad (6.19)$$

Provided the probe-to-sample separation remains constant (i.e. $C_{ps}$ does not change) and the sample is of uniform thickness ($C_S$ is fixed) the probe voltage is directly proportional to the surface charge density beneath it. Thus when the probe is scanned across the sample surface the output voltage is directly proportional to the local surface charge density.

In practice, the finite input resistance, $R_{pg}$, of the pre-amplifier imposes operating constraints. For a uniform surface charge density $\sigma_S$, the output voltage V does not in fact have the invariant value suggested by equation (6.19) but is an exponentially decaying function of time with a time constant $R_{pg}C_{pg}$ so that

$$V(t) = V_0 \exp\left(-\frac{t}{R_{pg}C_{pg}}\right) \qquad (6.20)$$

where t is the elapsed time from the start of the measurement and $V_0$ is the initial value measured. Consider the situation where a probe is scanned across a charged film sample in 10 s, and the allowed error in the measurement of $\sigma_S$ is not to exceed 1%, i.e.

$$1 - \exp\left(-\frac{t}{R_{pg}C_{pg}}\right) < 0.01 , \qquad (6.21)$$

so if $C_{pg} = 10$ pF, then $R_{pg} \geq 10^{14}\Omega$. Satisfying this input resistance constraint (inequality (6.21)) is not too difficult if the pre-amplifier has a MOSFET input stage. However, it is necessary to "zero" the probe potential regularly when the probe is positioned in a well-defined electrostatic environment. Typically this is carried out by placing the probe over a grounded metal strip which, in practice, may be one of the clamps holding the sample on the ground plane.

Careful attention has to be paid to probe/sample mounting and positioning mechanisms if reliable and consistent charge-density data are to be obtained. It has become customary to move the sample and its ground plane under a stationary probe, rather than vice versa, in order to avoid noise signals from flexing cables etc.

From equation (6.19) it will be obvious that for a given charge density the probe signal is maximised by making ($A\ C_{ps}$) as large as possible. The probe cross-sectional area (~A) is determined by the required resolution, while maximising $C_{ps}$ ($\sim\epsilon_0 A/d_{ps}$) requires $d_{ps}$ to be as small as possible. However, variations in probe-to-sample distance $d_{ps}$ give corresponding output voltage errors, so that a compromise has to be accepted, determined by the positional accuracy of the sample surface. In turn, the latter is dependent on uniformity of sample thickness and bearing tolerances on the moving sample holder.

Empirically, it has been found that resolution is ~1.5 times either the probe diameter or the probe-to-sample spacing, whichever is the greater. The optimal accuracy/resolution/signal-to-noise ratio compromise is achieved with $d_{ps}$ set equal to the probe diameter.

In practice, it is difficult to reduce $d_{ps}$ much below about 300μm. With a 300μm diameter probe positioned above a 100 μm thick sample having a surface charge density of 1μC/m$^2$, a signal voltage of about 2 mV is obtained. Nevertheless, probes as small as 80 μm have been developed (Baum et al, 1977; Gerhard-Multhaupt and Petry, 1983).

A number of distinctly different experimental arrangements have been developed to study charge distribution and migration. These have involved mounting the sample on a rotating cylinder (Zichy, 1970; Davies, 1970),a turning disc (Toomer and Lewis, 1979) or on a raster-scanned flat table (Hughes and Secker, 1971).

Figure 6.12 shows the pre-amplifier and zero-setting circuit for the last system, together with a plot of the charge density distribution on a charged polyethylene sheet. X and Y signals obtained from sample position sensors are used to drive an X-Y recorder and the charge density readout from the probe is fed in as a Y-modulation signal.

High resolution induction probes need to be maintained in a scrupulously clean condition to avoid performance degradation due to parasitic charge leakage. To avoid this constraint, and the difficulties of realising robust, drift free pre-amplifier designs, Nordhage and Bäckstrom (1976) developed a vibrating probe system in which the sensing element repeatedly passed across an aperture in a thin guard electrode close to the sample surface. The alternating signal from such a vibrating probe can be processed using a similar phase-sensitive detector to that used in field-mill systems (section 4.3.1). For the vibrating probe, the reference signal is derived from the circuit used to drive the probe oscillation device. Typically, a spatial resolution of about 1000 μm is achievable and, because of the continuous zero referencing, provides a much more robust measuring system than the induction probe. A range of instruments based on the vibrating probe technique is commercially available.

(a)

(b)

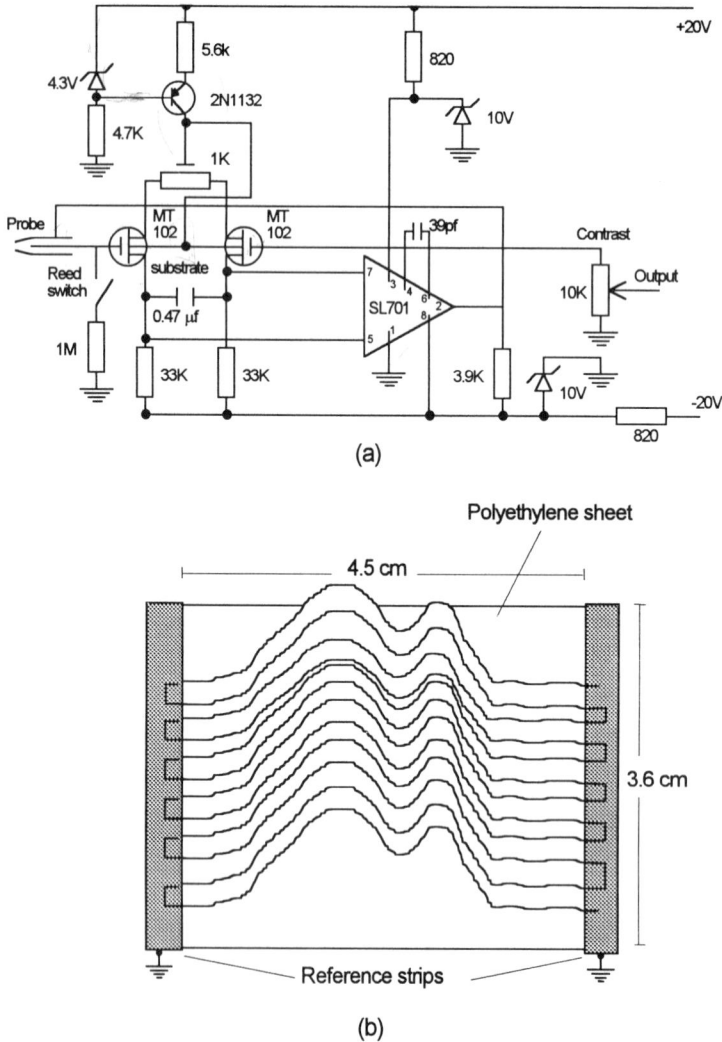

**Figure 6.12**    *Scanning induction probe showing (a) the preamplifier and zero-setting circuit and (b) a plot of surface charge distribution.*

It should be noted that,if the charge density on the sample surface is sufficiently great, a gas discharge may occur between the sample surface and the probe or its guard electrode. Such a discharge would certainly give rise to erroneous charge readings and may damage the probe preamplifier.

## 6.3 CHARGE DISTRIBUTED THROUGHOUT A VOLUME

For many electrostatic phenomena, charge is distributed throughout a volume of gas, liquid or solid. In most cases, measurement of charge density is effected by monitoring a related parameter such as the electric field at the boundary of the medium.

Suppose we consider a simple situation where a gas with a uniform charge density ρ per unit volume is contained within earthed conducting walls. The field at the gas/wall interface can be measured with an instrument such as a field mill set into the surface of the wall (see Figure 6.13). The relationship between the space charge density and the surface field is easily calculated using Gauss' theorem (section 2.3.4.1) as a starting point. Three elementary situations are considered:

(i)  A box-shaped container in which the height h (in the x-direction) is very much smaller than the length of the box.

(ii) A cylinder of radius R, where the length of the cylinder is very much greater than R.

(iii) A sphere of radius R.

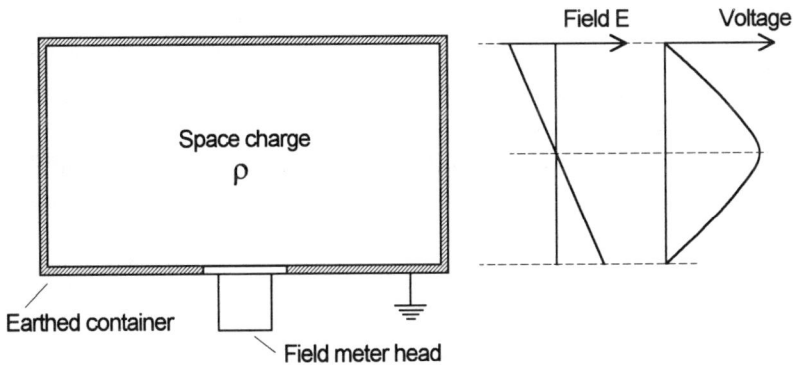

**Figure 6.13**    *Measuring the space charge field in a container. The field and potential distribution along the centreline of the tank are also shown.*

In each of these cases, the electric field has a maximum value at the gas/earthed wall interface, while the potential is a maximum at the centre of the container, where the field is zero. The expressions for calculating the respective field and potential values are set out in Table 6.1. For the three cases considered, it is interesting to estimate the magnitude of the field and potentials that may result from a low uniform charge distribution within a typical industrial-scale vessel. If we assume a charge density of $10^{-8}$ C/m$^3$, and the height, h, or diameter, 2R, is equal to 10 m, the resulting field and voltage values are as set out in Table 6.2. Of course, in practice, industrial vessels tend not to be of the simple elementary shapes described, but there is still a linear relationship between the volume charge density and the surface field, which can be established by computer modelling.

**Table 6.1**    *Field and potential maxima for various space-charge-filled vessels with conducting walls. The charge density has a uniform value Q in each case.*

| Vessel | Form of field variation | Field at wall | Potential at vessel centre |
|---|---|---|---|
| 1. Rectangular vessel height h, length and depth $>> h$. | $E(x) = \dfrac{Q}{2\varepsilon_0}(2x - h)$ | $\dfrac{Qh}{2\varepsilon_0}$ | $\dfrac{Qh^2}{8\varepsilon_0}$ |
| 2. Cylindrical vessel radius R, with length $>> R$ | $E(r) = \dfrac{Q}{2\varepsilon_0}r$ | $\dfrac{QR}{2\varepsilon_0}$ | $\dfrac{QR^2}{4\varepsilon_0}$ |
| 3. Spherical vessel, radius R | $E(r) = \dfrac{Q}{3\varepsilon_0}r$ | $\dfrac{QR}{3\varepsilon_0}$ | $\dfrac{QR^2}{6\varepsilon_0}$ |

As an alternative to undertaking computer modelling, a successful approach adopted in some cases is to surround the measuring field meter with an open weave mesh to define a quasi-earthed boundary zone of simple geometry. Typically, a mesh size of about 10 mm is used and the defining cage has a side length of at least 50 mm. Where the defined volume is a cube of side length c, it has been found (Hasse, 1977) that the measured field E is given to within 5% by the relation

$$E = \frac{Qc}{6\varepsilon_0}. \tag{6.22}$$

**Table 6.2**    *Maximum fields and voltages assuming Q = $10^{-8}$ C/m$^3$ and h or 2R = 10 m.*

| Vessel | Field at wall (kV/m) | Voltage at centre (kV) |
|---|---|---|
| Rectangular box | 5.64 | 14.1 |
| Cylindrical vessel | 2.82 | 7.1 |
| Spherical vessel | 1.88 | 4.7 |

This presupposes, though, that the charge flows sufficiently easily through the boundary mesh to ensure that the charge densities outside and inside are identical.

The fundamental problem in trying to monitor volume charge density is that a measurement of the field at the container wall merely gives a notional uniform volume charge density whereas, in reality, the charge may not be distributed uniformly.

### 6.3.1 Space-Charge Measurements in a Gaseous Medium

Where the charge density in a gas or a charged mist is to be measured, it is possible to incorporate a standard field mill in the container wall. Calculation of the space-charge/field relationship or calibration with a charged atmosphere of known charge density enables conversion ratios to be evaluated for the instrument.

Alternatively, the ions may be drawn by means of a small fan into an Ebert tube (Figure 6.14) formed from a pair of cylindrical coaxial conductors (Ebert, 1901; Chalmers, 1957). A voltage is applied to the inner conductor and a current measuring amplifier connected to the outer. The voltage is made sufficiently great that all the ions entering the tube are swept to the electrodes. Thus, the measured current is equal to the rate at which charge enters the tube, i.e.

$$I = \pi(b^2 - a^2)Qv \qquad (6.23)$$

where v is the gas flow velocity into the tube and $Q$ the charge density. For a sampling velocity of 2 m/s and an ion density of $10^{-8}$ C/m$^3$, a typical Ebert tube of effective diameter 5 cm would measure a current of $\sim 40$ pA.

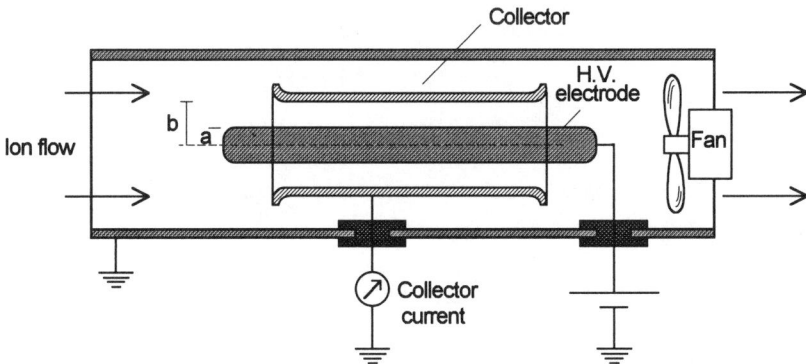

**Figure 6.14**   *Cross-section through an Ebert tube for measuring the ion density in a gas.*

Where electrostatic charge is held on mist or finely dispersed solid particles in a gaseous environment, the effective charge density can be measured by drawing samples through a filter made of conducting material. The current to ground from the filter element is monitored and the volume charge density evaluated from a knowledge of the gas volume throughput of the filter. This technique has been employed in assessing the degree of charging in mists created during the high-pressure washing of fuel cargo tanks in crude-oil tankers (Vos, 1971).

### 6.3.2 Space-Charge Measurement in Liquid

When a field mill is used to establish the volume charge density in a liquid of high resistivity, measurement errors can arise owing to the triboelectric charging associated with the relative movement of the rotor segments and the liquid. To reduce such errors field mills designed for use in liquids tend to have multivane rotors running at much lower speeds than is conventional for units operating in air. In such cases the field mill output voltage is a function of the signal frequency (see equation (4.16)) and hence the rotor must be driven by a constant-speed motor. An alternative solution to the problem (Chubb, 1983) involves modulating the local electric field at relatively low frequency with an earthed plunger oscillating through the sensor plate.

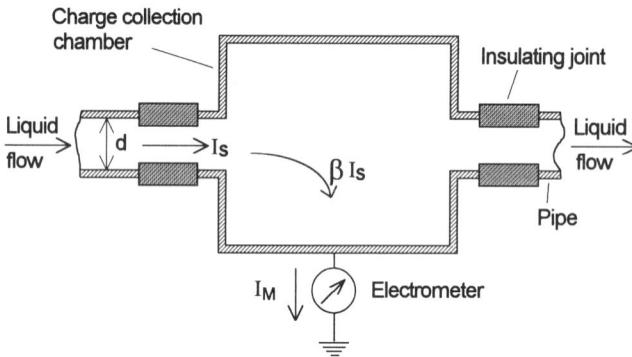

**Figure 6.15**    *Measuring the charge density in a liquid by collecting part of the streaming current in a relaxation chamber.*

A totally different approach to measuring the volume charge density in a liquid flowing through a pipe (Denbow and Bright, 1979) is illustrated in Figure 6.15. The large diameter cylindrical sampling section is insulated from the rest of the pipe through which liquid is flowing. The larger cross-section gives the liquid a significant residence time in the measurement chamber so that a major proportion of the space-charge will relax from the liquid under the action of its own space-charge field. A charged liquid flowing down a pipe gives rise to a streaming current, $I_S$ (section 3.1). For a pipe of diameter d with a uniform charge density $Q$ moving at a velocity v, the streaming current $I_S$ is given by

$$I_S = \frac{\pi d^2 Q v}{4}.$$  (6.24)

If the current $I_M$ collected by the measurement section is a proportion, $\beta$, of the streaming current then the charge density $Q$ can be calculated since

$$Q = \frac{4 I_M / \beta}{\pi d^2 v}.$$  (6.25)

$\beta$ is a function of (i) the geometry of the measurement zone and the liquid velocity both of which control the liquid residence time in the measurement zone and (ii) the relaxation time of charge in the liquid (section 2.9.4). Clearly, for operation with a given liquid, $\beta$ becomes just a function of liquid velocity - decreasing as v increases.

### 6.3.3   Volume Charge Density in Solid Materials

Solid insulators play an important part in electrostatic systems, but in most cases remain essentially uncharged, or only have charge distributed across external surfaces. This is because charge carriers do not move easily through insulating solids even when under high field conditions, and rapidly become immobilised in "traps" within the material. Thus measurements with insulating solids are more normally concerned with surface charge assessment than with attempting to measure a volume charge density. There are, however, a number of exceptions to this general situation, primarily associated with solids in relatively thin film form.

One situation where volume charge density and its spatial location need to be known are in thin film electrets used in microphone systems. The net charge within the film can be measured by placing it in a Faraday pail, while the depth below the surface of the charge layer can be calculated from careful measurements with the film lying on an earthed surface. Generally, this approach yields the magnitude and centroid of the charge.

Consider the arrangement shown in Figure 6.16 in which an insulating film with a charged plane lying a distance $\delta$ below its upper surface is placed between an earthed substrate and a field-measuring instrument.

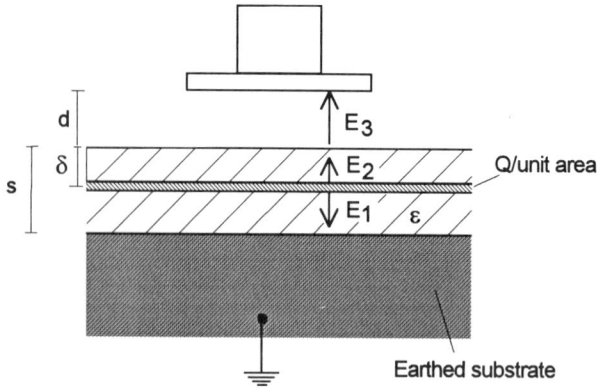

**Figure 6.16**    *Measuring the magnitude and depth of a charged layer in a solid.*

Application of Gauss' theorem leads to the equality

$$\varepsilon\varepsilon_0(E_1 + E_2) = Q. \tag{6.26}$$

The potential drop from the charge layer to both the earthed substrate and to the field-measuring instrument is identical so that

$$E_1(s - \delta) = E_2\delta + E_3 d. \tag{6.27}$$

At the sample/air interface, continuity of displacement requires that

$$\varepsilon\varepsilon_0 E_2 = \varepsilon_0 E_3. \tag{6.28}$$

Substitution between the preceding three equations leads to the equality

$$(s - \delta)\left(\frac{Q}{\varepsilon\varepsilon_0 E_3}\right) = \frac{s}{\varepsilon} + d. \tag{6.29}$$

Now we can see that if the spacing, d, between the sample and the field-measuring instrument is varied a plot of $(1/E_3)$ versus d should yield a linear graph from the the the slope of which $(s-\delta)$ or $\delta$ can be evaluated.

A much more direct method of measuring both the magnitude and distribution of charge within a film dielectric is to launch a short-duration pressure pulse normal to the film surface. As the pulse travels through the film at sonic speed, the temporary disturbance of any charge layers can be monitored by the resulting displacement current. One of the most effective methods of pulse generation is to use a high power laser to cause ablation of a graphite layer on the sample surface (Morisseau and Lewiner, 1987). The mechanical recoil resulting from the ejection of a few molecular layers of graphite results in a pressure pulse of about 1 ns duration. Typical displacement current traces are shown in Figure 6.17(b) for polymer samples that were electron-beam-charged at 20 kV and 40 kV (West et al, 1982). In both cases, there was an initial pulse corresponding to the compensating charge on the metal electrodes. The second pulse of opposite sign was related to the trapped electrons. Those injected at 40 kV were seen to be buried at about 14 $\mu$m below the electron-bombarded surface, while for the 20 kV injection the charge layer was some 5 $\mu$m below the surface.

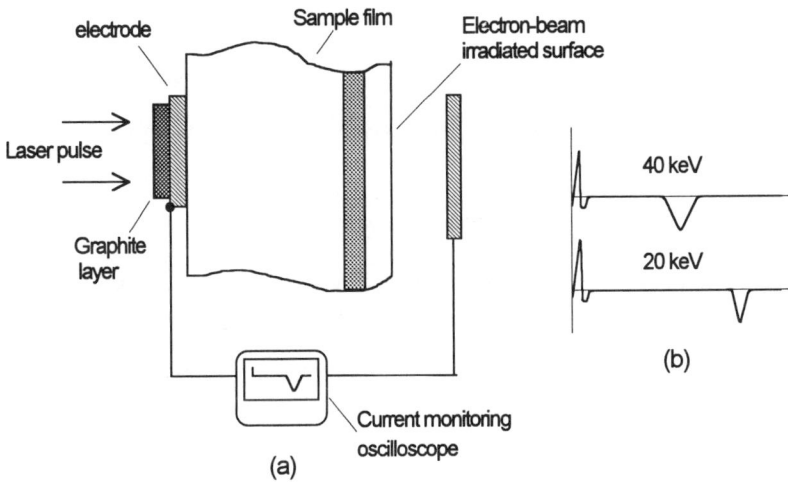

**Figure 6.17**    *(a) Laser-induced pressure pulse system for identifying a buried charge layer in an insulating sheet. (b) The signals recorded after irradiation with 40 keV and 20 keV electrons.*

## 6.4 CONCLUDING COMMENTS

Robust methods are available for measuring the total charge on objects of relatively small size, and surface charge density can be measured relatively easily. On the other hand, evaluating volume charge density in gases, liquids and solids presents many more problems. As we have seen, field measurements made at the earthed walls of containing vessels can give a notional value for the volume charge density assuming a uniform distribution of charge.

An assessment of the magnitude and distribution of charges in gaseous and liquid systems can be deduced by plotting the three-dimensional potential distribution within the system. A crude way of carrying out this task is to adjust the potential on a small corona probe until the current from it is zero (section 5.6). This then defines the local space charge potential (Wilson, 1970). It has also been found (Van der Weerd,1971) that an earthed field meter deployed within a charge-filled volume has a reading proportional to the local space potential (section 5.6). Thus, for large systems, a combination of relatively simple measurement techniques, i.e. field readings at the walls and potential readings throughout the volume, can give a fair estimate of both the magnitude of the volume charge as well as its spatial variation.

## REFERENCES

Baum EA, Lewis TJ and Toomer R 1977 "Decay of electrical charge on polymer films" *J Phys D: Appl Phys* **10** 487-497.

Chalmers JA 1957 *"Atmospheric Electricity"* (Oxford: Pergamon Press) 61-65.

Chubb JN 1983 "Developments in electrostatic fieldmeter instrumentation" *J Electrostatics* **14** 349-358.

Corbett RP and Bassett JD 1971 "Electric field measurements in ionic and particulate clouds" *Static Electrification 1971, IOP Conf Ser No 11* 307-319.

Cross JA 1987 *"Electrostatics: Principles,Problems and Applications"* (Bristol:Adam Hilger).

Davies DK 1970 "Charge generation on solids" *Advances in Static Electricity* **1** 10-21.

Denbow N and Bright AW 1979 "The design and performance of novel on-line electrostatic charge-density monitors, injectors and neutralisers for use in fuel systems" *Electrostatics'79 IOP Conf Ser No 48* 171-180.

Ebert H 1901 "Aspirationsapparat zur Bestimmung des Ionengehaltes des Atmosphare" *Phys Z* **2** 662-666.

Gerhard-Multhaupt R and Petry W 1983 "High resolution probing of surface charge distributions on electret samples" *J Phys E: Sci Instrum* **16** 418-420.

Hasse H 1977 *"Electrostatic Hazards"* (New York.:Verlag Chemie).

Hughes KA and Secker PE 1971 "A two-dimensional charge scanning instrument for flat insulating sheet" *J Phys E* **4** 362-365.

Morisseau D and Lewiner J 1987 "New possible measurements for the prevention of electrostatic discharges" *Electrostatics'87 IOP Conf Ser No 85* 197-202.

Moyle BD and Hughes JF 1982 "Particle charging and absolute measurement of charge-to-mass ratio" *Conf Record 1982 IEEE/IAS meeting San Fransisco Oct 4-7* 1191-1195.

Nordhage F and Bäckstrom G 1976 "Oscillating probe for charge density measurements" *J Electrostatics* **2** 91-95.

Toomer R and Lewis TJ 1979 "Charge effects at aluminium electrodes on insulating films" *Electrostatics'79 IOP Conf Ser No 48* 225-237.

Van der Weerd JM 1971 "Electrostatic charge generation during the washing of tanks with water sprays - II: Measurements and interpretation" *Static Electrification 1971 IOP Conf Ser No11* 158-177.

Vos B 1971 "Electrostatic charge generation during washing of tanks with water sprays" *Static Electrification 1971 IOP Conf Ser No 11* 184-92.

West JE, von Seggern H, Sessler GM and Gerhard R 1982 "The laser-induced-pressure-pulse method: a direct method of measuring spatial charge distributions in thin dielectric films" *Conf Record IEEE/IAS meeting San Francisco Oct 4-7* 1159-1161.

Wilson AD 1970 *PhD Thesis* Reading University.

Zichy EL 1970 "Electrostatic charges associated with dielectric surfaces" *Advances in Static Electricity* **1** 42-55.

# CHAPTER SEVEN

# RESISTANCE AND CHARGE DECAY

## 7.1 INTRODUCTION

In section 2.10 we showed that most electrostatic systems can be represented by the simple equivalent circuit in Figure 2.37. In this circuit, charge production is represented by the current generator. Charge storage occurs on the system capacitance while the resistance of the system allows the charge to dissipate. The magnitudes of the capacitance and resistance determine the decay-time (or relaxation time) of charge in the system (section 2.9.4), i.e. $\tau = RC$.

Where the production and retention of electrostatic charge form the basis of an industrial process, careful steps have to be taken to ensure that the magnitude of the resistance R remains high. This is necessary to ensure that the charge on which the required force effects are impressed is not prematurely neutralised or bled to ground. Conversely, the suppression of unwanted electrostatic side effects during processing requires R to be as small as possible. This necessitates the bonding to ground of both the process equipment and, as far as is possible, the process material itself. Clearly, then, the effectiveness of electrostatic processing as well as the avoidance of electrostatic hazard effects - fires, explosions, shock to operatives - are functions of the charge leakage characteristics of the process material or system.

For process systems, we are involved, primarily, with the integrity of the earth bonding of the constituent parts of the process machinery. For process materials, we are concerned with the volume or surface resistivity of the material. However, these parameters are not always particularly helpful since many materials of practical use demonstrate superlinear current-voltage characteristics indicating that the resistivity decreases with increasing applied electric field. Although national and international standards have previously categorised materials in terms of their volume and surface resistivities, there is an increasing recognition, at least for solid insulators, that a direct measurement of the charge relaxation time of the material may be a better indicator of its charge dissipative properties.

## 7.2  BONDING TO GROUND

To avoid electrostatic hazard all conducting elements of process plant must be firmly bonded to ground. If this precaution is not taken, the ungrounded plant element may store energy on its stray capacitance to ground by collecting charge from the material being processed. To avoid such a problem, making a connection to ground through a resistance less than 100 M$\Omega$ is normally capable of ensuring that charge accumulation and hence energy storage are negligible. Typically, a direct connection to ground would be made with the expectation of monitoring regularly a consistent plant-to-ground resistance of less than 10 $\Omega$.

Where process plant is built of semi-insulating or insulating material, e.g. wood or plastic, electrostatic effects are much more difficult to suppress. Except at very low relative humidity, wood has sufficient natural conductivity to bleed adventitious charge to ground. Plastic materials, however, can acquire charge very easily and may well retain that charge unless given an appropriate surface treatment. Volume additions to plastic materials have, in the main, been unsuccessful as so much additive is required to give a consistently high conductivity that the other physical properties of the plastic become totally changed - usually for the worse.

### 7.2.1  Practical Monitoring of Earth Bonding

Checking the bonding of process plant to ground is normally effected using standard resistance-measuring instruments. A difficulty occurs when the bonding link is, of necessity, highly resistive.

#### *7.2.1.1  Antistatic Boots*

An important example of a resistive earth bonding link is presented by process workers who must not be allowed to become charged because they work in flammable atmospheres or with static-sensitive electronic components. Despite the need to ensure that they do not retain charge, they must also be protected against serious shock injury in the event of accidentally coming into contact with the 250V mains supply. One solution to this problem is for the operatives to wear partially conducting footwear. Such footwear must connect its wearer to ground through an effective resistance between 50 k$\Omega$ and 50 M$\Omega$ (BS 5451 "Specification of electrically conducting and antistatic footwear"). The lower limit will ensure that if the operative inadvertently contacts the mains the resulting body current to earth is limited to a non-lethal value.

In practice it is found that leather-soled shoes meet the resistance specification above, but that shoes made of synthetic materials generally do not. Even those incorporating conducting additives can display widely varying resistance values. An additional problem for all types of shoe is that contaminant (insulating) layers collect on the bottom of the soles during normal use.

Sometimes the measured resistance is found to have a very definite voltage sensitivity. Measurement of effective operative resistance-to-ground therefore has to be undertaken with the operative simulating the situation against which the precautions are being taken. Figure 7.1 shows one instrument devised for this purpose. To test his footwear, the operative stands on the grounded plate and presses an operating button connected to a 500 V supply through a high value resistor $R_1$. The button is also connected to ground through another high value resistor $R_2$. When the operative presses the button, the button potential $V_b$ is thus determined by the resistance-to-ground of the operative. A high input impedance comparator circuit monitors the potential $V_b$, showing by means of output lights whether the operative's boots are providing effective bonding to ground.

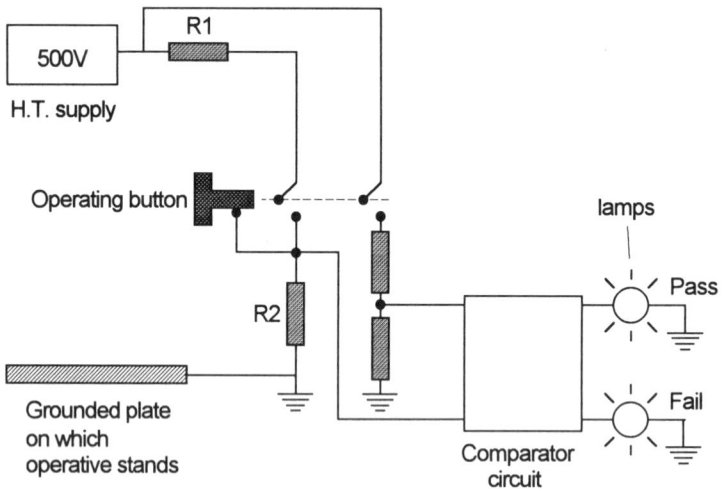

**Figure 7.1**    *A footwear resistance ground monitor.*

An identical circuit can be used to test the effectiveness of wrist straps worn by personnel involved in assembling and handling static-sensitive electronic devices and circuit boards.

### 7.2.1.2  Buried Conductors

In certain cases, it is necessary to confirm that earth bonding is in place, even though direct access to the bond system is not possible. Such a situation arises with aircraft radomes, where the basic insulating structure is painted with a partially conducting layer connected to the frame of the aircraft. This paint film minimises signal distortion on the underlying radar antennae caused by tribocharging/surface discharges on the radome surface. To protect the partially conducting layer from abrasion by ice-particle impact,it, in turn, is covered by a further insulating coating.

Taillet (1979) proposed a method of measuring the ground connection using an alternating signal to give capacitive coupling through the protective coating to the underlying partially conducting paint film. Figure 7.2 shows the equivalent circuit associated with the measurement. The effect of the coupling capacitance between the probe and the underlying partially conductive film can be minimised by making the probe area relatively large, and operating the system at a sensibly high frequency.

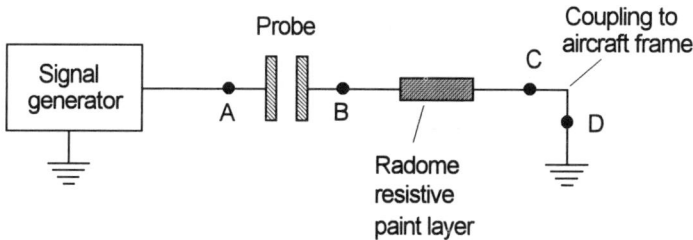

**Figure 7.2**     *Equivalent circuit of a radome tester.*

In one implementation (Figure 7.3) of such a measuring system (Secker, 1980), a 20 kHz signal source was used in combination with a 10 cm diameter probe. The probe surface was covered with a conducting foam to ensure good contact to the radome surface. In this case, if the protective insulating layer is 0.4 mm thick, the impedance of the probe-to-paint film capacitance is of the order of 20 k$\Omega$, which is negligible compared with that of the partially conducting paint film. Obviously, by adding a phase sensitive detector, it is possible to select unambiguously the reactive and resistive elements of the circuit between points A and D shown in Figure 7.2.

There is a slight concern with this approach, in that the integrity of the connection to local earth is not fully monitored. Consider the case where the connection between a radome resistive film and the aircraft frame is made by six bolts and washers with a total contact area of 30 cm$^2$. If corrosion develops on the washers such that a very thin insulating surface 0.02 mm thick develops on the washer surface, an additional capacitance is interposed between C and D in the equivalent circuit in Figure 7.2. This capacitance has a value nearly an order of magnitude greater than that of the probe-to-paint film capacitor, i.e. the capacitor between A and B. Thus, although the measuring technique can readily monitor the resistance of the surface film, it does not provide totally reliable confirmation that the film itself has a high integrity connection to ground.

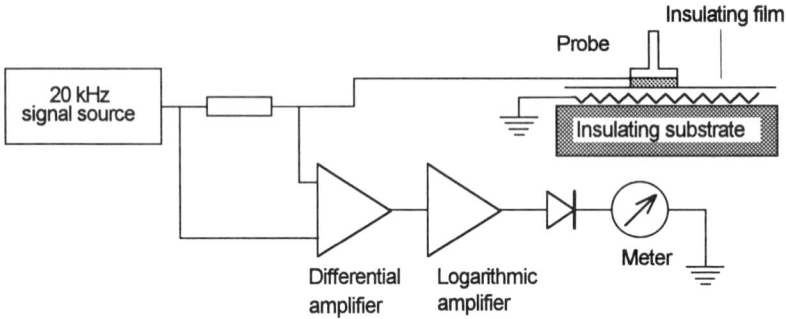

**Figure 7.3**    *Non-contact resistive film monitor.*

### 7.2.1.3  Work-Pieces on Conveyors

An entirely different check has been developed for the ground connection in electrostatic painting and powder coating systems, where it is important that the work-pieces are securely earthed to ground through the supporting conveyor on their passage through the coating booth. Due to the repeated cycling through the booth, an insulating paint or powder layer can easily build up on the support hooks such that electrical contact between the work-piece and the earthed hook is inhibited. Failure to provide such earthing results, at minimum, in very poor coverage of the work-piece by paint or powder because of the work-piece charging up during the coating operation. There is also the possibility of flashover between the work-piece and its supporting hook, with the attendant ignition risk. In addition, operatives off-loading the product may receive a shock.

Figure 7.4 shows a pre-booth monitoring system (Secker, 1979) for confirming that every work-piece is satisfactorily grounded. This operates without making direct physical contact with the work-piece; in a practical industrial system, such a contact would need to be so robust that the resulting surface damage to the work-piece would be unacceptable, particularly for "white goods" for which electrostatic painting and powder coating are very popular.

The operation of the monitoring system is extremely simple in concept. The upstream corona source attempts to charge up each work-piece. If its resistance to ground via the support hook is greater than $10^8$ $\Omega$ there will be significant residual charge on the work-piece when it passes the field mill downstream of the corona source. The output from the field-mill thus provides a signal which can be used to trigger appropriate alarm systems and to initiate a shut-down of the coating line.

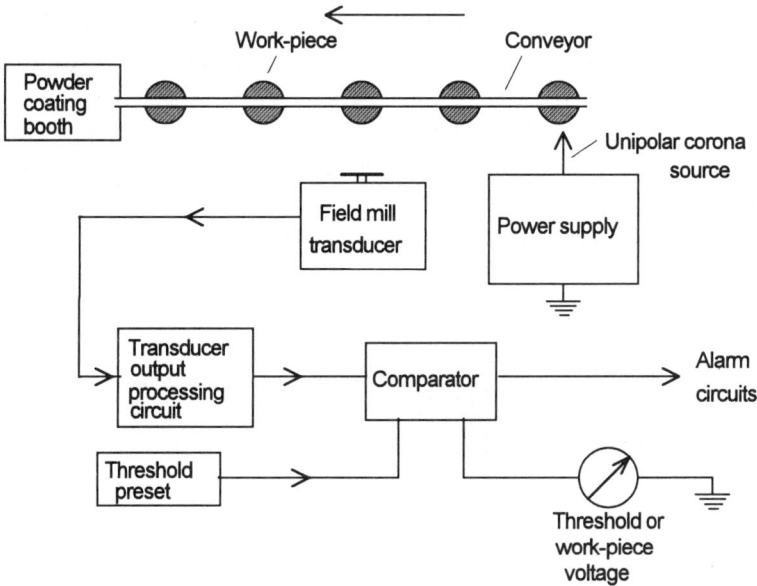

**Figure 7.4**    *A non-contacting detection system for spray-booth work-piece grounding.*

## 7.3  CHARACTERISATION OF PROCESS MATERIALS

We find,in many cases, that it is important to measure the resistivity of process materials prior to use, either to ensure that the value is sufficiently low to quench undesired electrostatic effects, or, alternatively, that it is sufficiently high so that desired electrostatic phenomena are not inhibited.

### 7.3.1  Volume Resistivity

A volume resistivity measurement is carried out by measuring the electrical resistance presented by a sample of the material when sandwiched between two electrodes of well-defined geometry (section 2.9.2 and Figure 2.32).The resistance so measured is converted to a resistivity through equation (2.55). The detailed manner of executing the measurement depends on the form of the material and in particular on whether the process material is a solid, liquid or powder.

### 7.3.1.1   Volume Resistivity of Solids

It is common practice, though not to be recommended, for the electrodes to be painted onto the sample surfaces with conducting silver paint or colloidal graphite. While this may be convenient, the risk of obtaining erroneous results is increased since solvent from the paint may diffuse into the material, carrying the conducting paint particles with it. It is preferable to vacuum evaporate or to sputter thin films of gold or aluminium onto the sample surfaces through a mask. Alternatively, the sample may be clamped between two flat, highly polished, stainless steel electrodes one of which, ideally, is sprung-loaded to effect a uniform, well-defined contact pressure. For materials with large volume resistivities it is advisable to use a guard ring around the current measuring electrode (Figure 7.5) to prevent surface leakage currents from being measured and to minimise the distortion of the electric field in the sample.

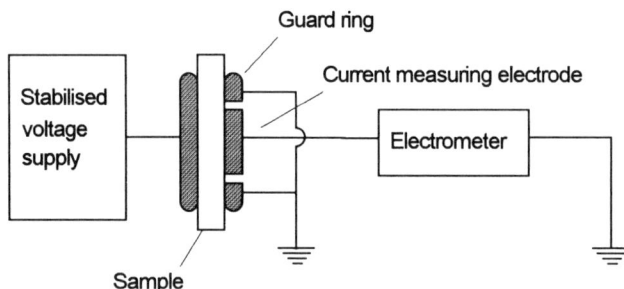

**Figure 7.5**    *Measuring volume resistivity using an electrode system with a guard ring.*

When the measured current is less than about 1 nA it will be necessary to ensure that the sample and measuring system are electrically screened to reduce disturbances from electrical noise. A stable voltage source is also required otherwise noise from the voltage source will capacitively couple into the current measuring instrument which may be a commercial electrometer or a virtual earth amplifier furnished with an appropriate feedback resistor (section 4.2).

Currents as low as 0.1 pA can be measured with a modest amount of care. For safety reasons it is usually advisable to apply a voltage no greater than, say, 100 V across the sample. Assuming the area of the current measuring electrode in Figure 7.5 is 10 cm$^2$ and that the sample is 0.1 mm thick, then from equation (2.55) the maximum volume resistivity, $\rho$, that can be measured is

$$\rho = \frac{V}{I} \times \frac{A}{d} = \frac{100}{0.1 \times 10^{-12}} \times \frac{10 \times 10^{-4}}{0.1 \times 10^{-3}} = 10^{16} \Omega m \ . \qquad (7.1)$$

When necessary, and with appropriate safety considerations, the supply voltage can be increased to 10 kV in order to extend the range of measurement to higher resistivities or to thicker samples, or even to allow a reduction in the sensitivity of the current measuring instrument required for the measurement.

### 7.3.1.2 *Volume Resistivity of Liquids*

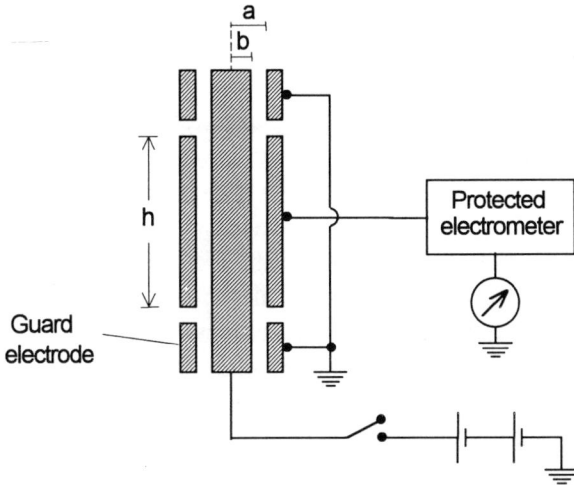

**Figure 7.6**     *Cylindrical test cell for measuring liquid resistivity.*

The volume resistivity of liquids is measured in a suitable conductivity cell. Such cells generally consist of cylindrical, coaxial electrodes between which there is a relatively small gap. To minimise end effects, cylindrical guard electrodes are also provided. Figure 7.6 shows a typical measuring system. If a and b are the radii of the outer and inner cylinders respectively, h is the effective length of the measurement zone and a current I is measured for an applied voltage V, the resistivity of the liquid is given by the expression

$$\rho = \frac{V}{I} \times \frac{2\pi h}{\ln{(a/b)}} \qquad (7.2)$$

If the liquid exhibits non-ohmic behaviour, it is important to make the measurement with the field in the liquid similar to that of the application for which it will be used. Since the gap (a-b) is very much smaller than either a or b, the measurement field is given approximately by

$$E = \frac{V}{(a-b)} \ .$$

(7.3)

In a typical measurement cell (a-b) is 1 mm and measurements are made with applied voltages of between 100 V and 1 kV giving interelectrode fields in the range $10^5$ to $10^6$ V/m.

### 7.3.1.3   Volume Resistivity of Powders

Organic powders are either the finished product or the intermediate output of many industrial processes. They are also widely used in coating applications. Many of these powders exhibit very high resistivities, the particular value indicated by measurement often being more associated with trace impurities or the humidity of the environment rather than the fundamental resistivity of the material itself. As in the case of liquids, powder resistivity is measured in a cell furnished with guard electrodes.  In this case though, it has become conventional to use a parallel-plane electrode system such as that shown in Figure 7.7. This is partly because it is difficult ensuring that the powder sample fills the cell uniformly - it is easier to fill a planar cell than a cylindrical one. Consistency of cell filling obviously affects the results, but it is now generally accepted that "firming" the sample by tapping the filled or partially filled cell on the bench gives as good performance as is obtained by more sophisticated pressurising or mechanical compacting systems.

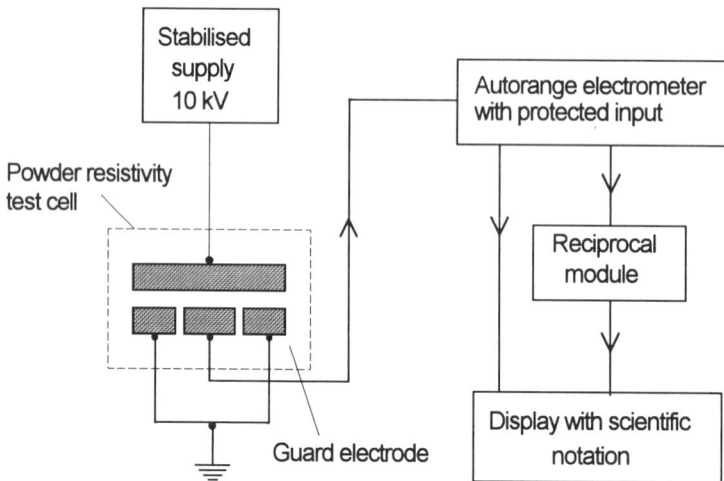

**Figure 7.7**    *Powder resistivity measurement system.*

With resistivity values which may be as high as $10^{16}$ $\Omega$m, powders pose some interesting resistance-measuring problems. In one commercially-available system (Secker, 1980) the current measuring electrode has an effective area of 5 cm² and the inter-electrode gap is 5 mm. With 10 kV applied across the electrodes, the current could be as low as 0.1 pA. With such low currents to be measured, the need for (a) guard electrodes and (b) cell construction from a good insulator such as PTFE becomes obvious.

While measuring the current is not in itself a major difficulty, we can see that this particular application does give rise to special concerns. If, due to bad filling, air pockets are trapped in the powder between the electrodes, or there are conducting impurities in the sample, a sparkover is likely to occur on applying the 10 kV test voltage (which provides an interelectrode stress of 2 MV/m). It is essential, therefore, to incorporate fast-acting protection circuits in the current-measuring system.

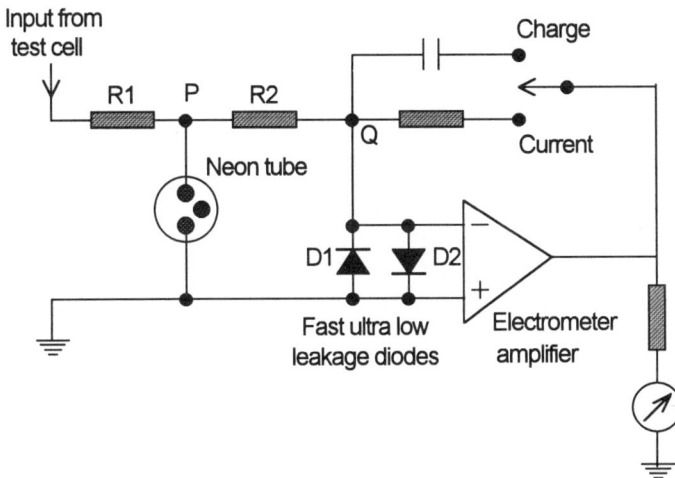

**Figure 7.8**    *A two-stage electrometer protection circuit.*

Figure 7.8 shows a typical protection circuit for an electrometer operating in virtual earth mode. Normally, point Q is held very close to earth potential and the low-leakage diodes $D_1$ and $D_2$ are non-conducting. If a high-voltage transient appears at the input to the measuring system due to sparkover in the test cell, the neon tube breaks down holding the transient voltage at point P to about 200 V with $R_1$ chosen to restrict the neon tube current to a safe value. The potential at Q may depart from zero, but diodes $D_1$ and $D_2$ hold the perturbation between - 0.8

and + 0.8 V, so that the electrometer amplifier is not damaged. The fault current through $D_1/D_2$ is limited by $R_2$.

Experience shows that measurements of powder resistivity may be quite time-consuming, as significant "settling" periods may be required for stable sample currents to be observed. Nevertheless, in processes such as powder coating it is useful to be able to obtain a direct "indication" of resistivity to ensure that moisture adsorbed from the atmosphere has not seriously lowered the effective resistivity.

### 7.3.1.4 *General Considerations*

Static-related problems with process materials are unlikely to occur when their volume resistivity is less than about $10^8$ $\Omega$m, while for applications such as powder coating, volume resistivities as great as $10^{16}$ $\Omega$m may need to be achieved in order to maximise the charge retention properties of the material.

For most materials within this range of interest, the resistivity, $\rho$, is a strong function of temperature, the dependence being given by an exponential relationship of the Arrhenius type, namely

$$\rho = \rho_\infty \exp\left(\frac{W}{kT}\right) \qquad (7.4)$$

where $\rho_\infty$ is a constant, W an activation energy, k is Boltzmann's constant and T the absolute temperature. Consequently, we find that the resistivity of a material with an activation energy of 1 eV (ca. 100 kJ/mole) will decrease by about 14% per °C increase in temperature around room temperature. Thus, if a particular production step requires the material to be heated, the resistivity of the product could change by several orders of magnitude as it passes through different stages in the production cycle.

The situation is further complicated because the activation energy, W, may well be dependent on the magnitude of the electric field applied across the material, W decreasing as the field increases. This is a manifestation of the fact that an electric field increases both the generation rate of free charge carriers in the material and their mobility.

When a voltage is applied across samples of certain insulators, particularly those with resistivities above about $10^{10}$ $\Omega$m, the current through them decreases with time. For some insulators the current can decrease by orders of magnitude over many hours. This process is known as polarisation and may have several different origins, e.g. exhaustion of mobile charge carriers, limited dipolar response, electrode effects, charge trapping and so on. The presence of such processes results in the apparent resistivity of the material increasing with time. The question then arises as to the true value of the sample resistivity. Conditioning procedures for overcoming some of these effects have been reported in the literature but these may only be relevant to the specific materials for which the procedures were evolved.

As a general rule, the value of resistivity measured several minutes after applying a voltage will give a reasonable guide as to the static retaining properties of the material. However, the measurement must be carried out at the same electric field(s) and temperature(s) that the material will meet during processing and/or subsequent application.

### 7.3.2   Surface Resistivity

It is more usual to characterise some film products and laminates in terms of their surface resistivity, $R_S$. This parameter is also used in many international standards to define materials for use in various hazardous environments. As seen in section 2.9.3 and Figure 2.33, the measurement can be carried out by recording the current which flows between two linear, surface mounted, coplanar electrodes following the application of a test voltage. In such a measurement it is assumed that the current flows from one electrode to another through a thin surface skin.

Better screening of the sample and system electronics is possible by using the concentric ring or concentric cylinder arrangements shown in Figure 7.9. With the concentric-ring electrode structure, particular care must be taken over the choice of substrate on which the sample sits. If the electrical stress is to be primarily along the sample surface, the sample support must be as near as possible to an ideal insulator. The use of an earthed conducting substrate encourages bulk conduction. In fact, whether there is surface conduction or a mixture of surface and bulk conduction depends primarily on the physical state of the surface and its ability to support conduction mechanisms.

Recent three-dimensional field calculations (Leonidopoulos, 1991) have shown the superiority of the concentric-cylinder arrangement (Figure 7.9(b)) over the concentric-ring configuration (Figure 7.9(a)). Not only does the former give enhanced shielding, but the electric field adjacent to the electrodes is lower than with the ring electrodes, thus reducing the probability of charge injection into the test sample.

With cylindrical electrodes, the measured resistance R is converted to surface resistivity, $R_S$, using

$$R_S = \frac{2\pi R}{\ln (a/b)} \qquad (7.5)$$

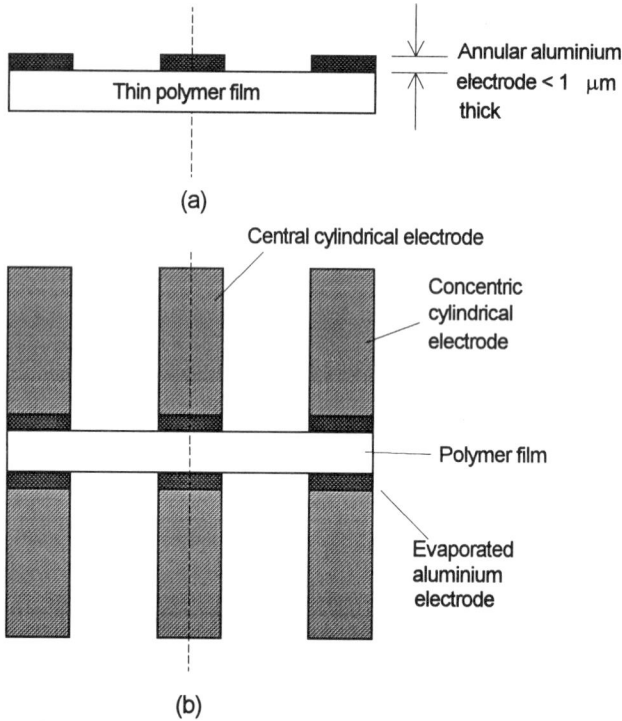

**Figure 7.9**    *(a) Concentric-ring and (b) concentric-cylinder electrode configurations for surface resistivity measurement.*

where a and b are the radii of the outer and inner cylinders respectively.

### 7.3.2.1    General Considerations

Polarisation processes similar to those which cause the volume resistivity of insulators to increase with time of application of the test voltage will also be active in measurements of surface resistivity. These effects usually manifest themselves in materials whose surface resistivities are in excess of about $10^8$ $\Omega$ per square. The surface resistivity will also follow an Arrhenius law and will therefore be sensitive to changes in ambient temperature.

It is important to realise that the surface resistivity of many nonconductors is very sensitive to the ambient relative humidity. In fact, certain types of antistatic agents are applied to the insulator surface precisely to encourage the adsorption of a layer of water to the surface. Such materials have low surface resistivities at a relative humidity of say 60%, but become highly resistive when the humidity falls below 30%. At these latter humidities the moisture content of the air is so low that insufficient water molecules remain on the surface to form a continuously conductive layer. Consequently, it is important that at least one measurement of surface resistivity is carried out at the lowest relative humidity that the material is likely to encounter in use. It is important,that prior to any measurement, the sample should be conditioned for several hours at this humidity value.

### 7.3.3  Charge Decay Measurements

The difficulties inherent in making and interpreting resistivity measurements, coupled with the non-ohmic behaviour of most insulating and semi-insulating materials,have made the direct measurement of charge decay rates from materials a much more attractive option for characterising their charge dissipation properties. Furthermore, the charge decay rate (or relaxation time) is much more readily compared with process speed so that a judgement on the likely charging effects of the material in the process can be reached more easily.

#### 7.3.3.1  *Liquids*

One of the earliest uses of the charge decay technique was described in detail by Klinkenberg (1958) for characterising liquid petroleum products. The concept of the measurement can be understood with the aid of Figure 7.10. The test cell consists of a conducting container of about 1 litre capacity with a spherical electrode suspended from its insulating lid. The sphere is connected to an

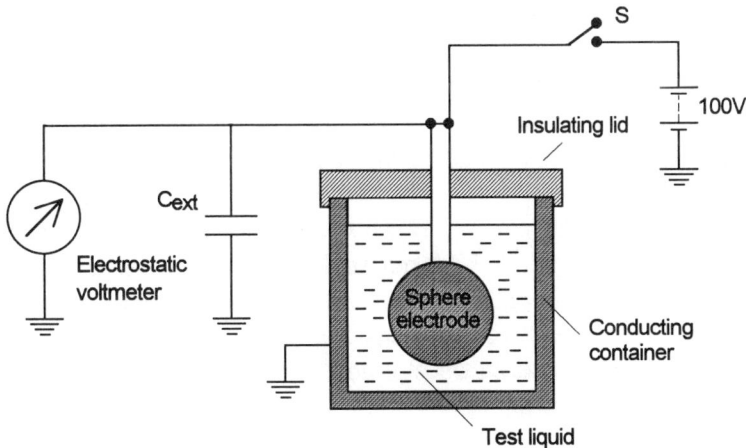

**Figure 7.10**    *Charge leakage measurement in liquids.*

electrostatic voltmeter and, if necessary, additional external capacitance can be added to the system. The sphere is charged by closing switch S, and the voltmeter reading is monitored after the switch is opened. As charge leaks from the sphere by conduction through the liquid, the observed voltage decays with time t following the exponential law

$$V = V_0 \exp\left(-\frac{t}{RC}\right)$$                     (7.6)

where R is the resistance of the liquid sample and C is the sum of the cell capacitance and the external capacitance, $C_{ext}$. If $C_{ext}$ is small compared with the cell capacitance then the RC time constant will be approximately equal to the relaxation time of charge in the liquid (section 2.9.4).

In making measurements of this kind it is obviously necessary to ensure that the insulation of the test cell itself is extremely good and that the leakage of any external capacitors is very low, such that the measured voltage decay with the test cell empty is very much slower than when the sample is present.

### 7.3.3.2  Films and Laminates

There are many examples in industry where solid materials, in film, sheet or fabric form, are used in a support or packaging role, or are incorporated into bench tops, flooring, upholstery or clothing. In situations where there is a sensitivity to charging (for example, processing in a flammable environment, transporting electrostatic-sensitive devices) it is important to check that any charge generation, typically resulting from tribocharging, is quickly bled to ground.

Figure 7.11 shows a very simple test rig for monitoring the charge leakage behaviour of sheet/film material. The material is held in an earthed conducting frame and charged by rubbing it vigorously with a second material (one that is well-separated in the triboelectric series from the material under test). The subsequent charge decay is monitored with a field mill. An exponentially decreasing field should be monitored as charge bleeds to the supporting frame. The time constant for charge decay may be defined as the time for the field to decrease to 1/e,i.e. 37% of its original value. Alternative definitions often used are (i) the time for the field to decrease to half its original value - the half-charge decay-time, and (ii) the time to decay to 10% of its original value.

In practice, the above measurement is not very satisfactory since tribocharging gives very inconsistent performance, and for those interesting materials with short decay-times there are difficulties in disentangling the charging and the charge decay actions. To some extent this problem can be overcome by changing the charging method.

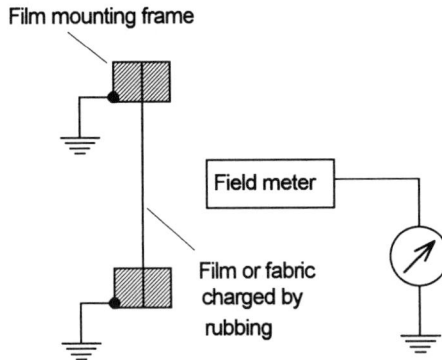

**Figure 7.11**    *Basic arrangement for measuring surface charge decay from homogeneous films and fabrics.*

In one embodiment (US Federal Standard 101C, Test Method 4046), the conducting frame holding the sample is connected to a voltage supply and the field from the sample measured as a function of time. When the field attains its maximum value the frame can be earthed and the decay of field with time noted. The rise- and decay-time should be the same. The method gives reliable results for *homogeneous* materials with uniform volume resistivity and for materials with a uniform surface resistivity arising, say, from surface-applied antistatic agents. However, care must be taken to ensure that the fieldmeter sensing head is electrically screened from the conducting frame to which the voltage is applied.

The voltage charging technique described above is not recommended where materials are not homogeneous. In particular, the method is unsuitable for those materials which rely on the presence of conducting fibres for the suppression of electrostatic effects. For such materials the test will only provide data on the rapid transport of charge along the conducting fibres during the charging step and its dissipation from those same fibres after earthing. No information will be gleaned concerning the charge retention properties of the insulating matrix in which the conducting fibres are embedded. Thus, charges generated triboelectrically on the insulating matrix may well have a considerably longer decay time than indicated by the test.

A commercial instrument for measuring charge decay from homogeneous materials has been described by Taylor et al (1987). As can be seen from Figure 7.12 the principle of operation is similar to the Klinkenberg method for liquids. An isolated electrode resting on the sample surface is charged by momentary connection to a voltage supply. As charge leaks to the earthed sample clamp, either through the sample or along its surface, the potential of the electrode decays. The rate of decay is monitored by a fieldmill and facilities exist for recording the whole of the decay curve and for extracting the decay time automatically.

There can be no doubt that the surface charge distribution in this measurement will differ from that which occurs in practice with tribocharging. For homogeneous materials, including laminated structures formed from different but homogeneous materials,this is not a serious problem. Composites formed from highly insulating materials containing conducting fibres again pose special difficulties, since the highly insulating regions play little role in the above measurement. In practice, it is on just such regions as these that tribocharge will be generated and held for long periods of time. There is concern, therefore, that measurements made by this method may flatter the charge-dissipative behaviour of the test material.

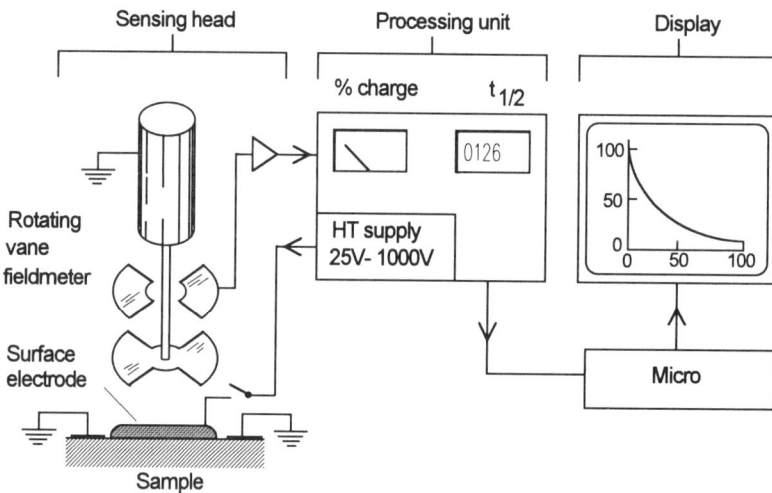

**Figure 7.12**    *A commercial charge decay meter.*

To overcome this objection Wilson (1990) has suggested a modification to Federal Test 101C so that sample charging is effected with a corona discharge. During the charging stage the sample frame is disconnected from ground and measurement of the decay-time commences from the instant the frame is earthed. Of course, there will be an immediate and rapid fall in the measured field due to an electrostatic screening effect. This is caused by the flow of induced charges from the earthed frame along the conducting fibres. These induced charges are of opposite sign to the corona charge and effectively neutralise them even though direct recombination does not occur. The magnitude of the initial decrease in field will depend on the average separation of the fibres and on the distance of the fieldmeter from the sample surface. Closely spaced fibres will screen more effectively than widely spaced ones. The attenuation of the field at the field meter will be greater the further away it is from the sample surface. Consequently, care is required in how the charge decay time is defined. The useful information concerning the insulating matrix will be contained in the "tail" of the field vs. time plot and may constitute only a few percent of the initial field.

In the commercial measurement system (Figure 7.13) developed by Chubb (1988),charging of the sample surface is achieved by a short burst of corona ions from a multi-point corona source positioned between the surface and field mill measuring instrument; the corona system is then removed very rapidly from the measurement zone using a mechanically-sprung arrangement and the surface charge decay monitored. In this system, though, the sample perimeter is grounded throughout the charging step which means that induced charges will already be present on the conducting matrix when monitoring of the charge decay process commences. Thus, only the behaviour of the "insulating" regions is recorded.

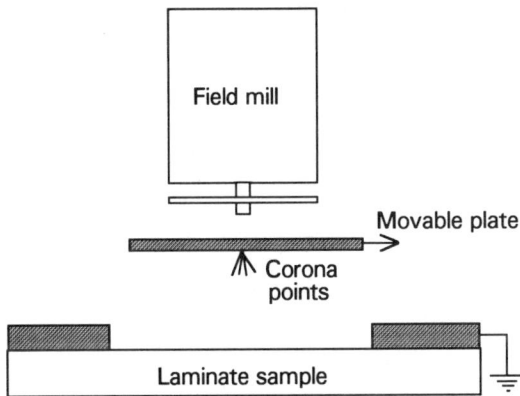

**Figure 7.13**   *Method for corona charging a surface prior to a charge decay measurement.*

Provided that the charging time is short compared with the charge decay-time constant, the charge distribution at the start of the charge decay measurement will corrrespond to that deposited from the corona source. This charge distribution is essentially determined by the geometry of the corona source and is not greatly affected by the resistive homogeneity of the sample surface, providing it does not contain ground-connected fibres. Thus, very highly insulating regions in an other- wise uniform high resistivity matrix will become charged and ,for composite samples, a quasi-uniform initial surface charge density should be achieved.

We should appreciate, however, that the charge species on the surface will almost certainly be different when charging is effected by a corona source rather than by tribocharging. It is possible also that the surface properties may even be modified as a result of bombardment by ions and by the chemically-active, excited neutral species generated in the corona. Furthermore, corona ions are readily hydrated and may carry moisture to the surface under test. This may be an important factor since the corona ions will be deposited over most of the sample surface.

The choice of test method for measuring the charge decay-time will depend very much on the nature of the application to which the material is to be put. For homogeneous materials, and laminates constructed from homogeneous materials, the different test methods should give similar values for decay-time assuming that the system capacitance, sample geometry and initial surface voltage are the same in all cases. When dealing with composite materials containing conducting fibres we must distinguish two possible cases:

(i) If the material is to be used for dissipating static charge from charged bodies placed on it, then the charged electrode approach in Figure 7.12 is an appropriate test.

(ii) If there is concern that charge generated on the insulating regions of the composite material itself may pose a problem, then corona charging is the more appropriate approach.

Charge decay meters reliant on a field mill to measure the charge decay will obviously be limited by the response time of the instrument which, typically, is about 25 ms. For more rapid decay rates, the stator plates of the field mill can be used as an induction probe, enabling fast transients to be followed (Taylor and Elias, 1987). In the corona-charged system a lower limit of 50 ms for the charge-decay time constant is set by the time taken for the corona source to be moved from the measurement zone on completion of the charging phase.

### 7.3.3.3  On-Line Surface-Charge-Decay Gauge

Where highly insulating films need surface treatment to ensure that satisfactorily-fast charge decay occurs, it is clearly desirable to build in a measuring system which continuously monitors the charge-decay characteristic realised by the treatment process.

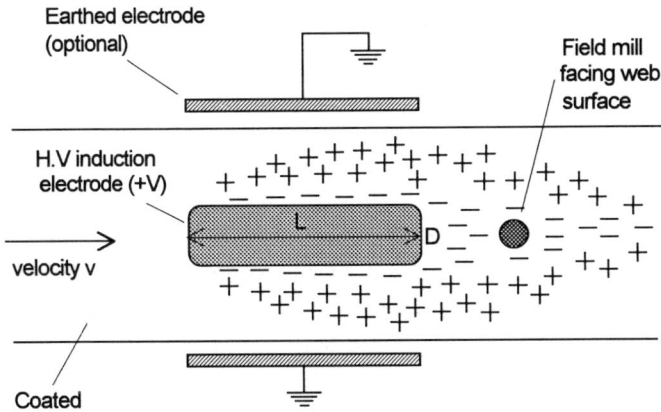

**Figure 7.14** *Positively charged induction electrode attracts negative charges to central section of web/film but repels positive charges to the edges. Downstream of the electrode the charges relax..*

Figure 7.14 shows the basis of an on-line, non-contact measurement in which a high voltage electrode causes charge separation on the film surface as a result of the transverse field. Normally, the earthed frame of the film support system will provide a low-potential environment beyond the film edge, but earthed electrodes can be fitted if necessary. Once the film has moved downstream of the high voltage electrode, recombination of the separated charges takes place at a rate determined by the charge decay time constant.

At the downstream end of the induction electrode, the surface charge density, $\sigma$, is given by the expression

$$\sigma = \sigma_0 \left\{ 1 - \exp\left(-\frac{L}{v\tau}\right) \right\} \qquad (7.7)$$

where $\sigma_0$ ($= \varepsilon_0 E_0$) is simply related to the field $E_0$ established at the film surface by the electrode, $v$ the linear velocity of the film and $\tau$ is the charge-decay time constant.

The film takes a time $x/v$ to move from the induction electrode (point D) to the field mill measurement position. In this time, charge recombination takes place and the residual charge density $\sigma_{FM}$ at the field mill is given by

$$\sigma_{FM} = \sigma_0 \left\{ 1 - \exp\left(-\frac{L}{v\tau}\right) \right\} \exp\left(-\frac{x}{v\tau}\right). \qquad (7.8)$$

Thus, monitoring the field mill output provides an indication of the variation of the charge decay-time of the section of film moving under the measurement system. Blythe (1983) combined such a system with a traversing mechanism that enabled the induction electrode/field mill to be scanned over the width of the film, and with this he was able to detect non-homogeneity in the charge-decay characteristics of the surface coating on the film.

In practice, the simple system described above has a number of aberrations, in that the field measured will be influenced by any initial charge on the film upstream of the induction electrode and will also contain a component arising from the fringe field at the leading and trailing edges of the induction electrode. Furthermore, if separation between the film and the field mill varies due to variations in film tension, the measured field will change. These difficulties can be partly allieviated by introducing a second field mill downstream of the first. The two field mill readings, $E_1$ and $E_2$,should be related by the expression

$$\frac{E_1}{E_2} = \exp\left(\frac{x_2 - x_1}{v\tau}\right)$$  (7.9)

Rearranging gives the decay time-constant, $\tau$, directly,i.e.

$$\tau = \frac{x_2 - x_1}{v \ln(E_1/E_2)}$$  (7.10)

Chubb et al (1991) describe an on-line system with three field mills positioned 14, 39 and 114 mm downstream of the induction electrode. By selecting readings from appropriate pairs of field mills, it is possible, under microprocessor control, to cater for quite a wide range of charge decay time-constants, as well as providing alarm and line shut-down signals in the event of the processed film not having an adequate charge leakage capability.

### 7.3.3.4  *General Considerations*

Experience indicates (Taylor et al, 1987)  that charge-decay curves do not follow the single exponential characteristic that simple theory would lead us to expect. Typically, an initial rapid decay of charge is followed by a much slower "tail".  This gives the clue that charge decay is a multi-phenomenon process, involving electronic and ionic conduction, charge trapping and field-dependent polarisation effects. Obviously, where the sample is not homogeneous, interface effects between the differing micro-regions provide other opportunities for complicating the charge-decay mechanisms.

We should note also that for many "antistatic" film materials, the observed charge-decay characteristics are a consequence of a surface treatment (or volume addition to the batch material) which results in moisture from the atmosphere being trapped on the surface. As with the surface resistivity measurement, it is important to ensure that the test sample is held in a constant humidity environment prior to the measurement and that this humidity should represent the most severe operating environment (i.e. the lowest humidity) to which it will be subjected. To ensure that satisfactory performance will be achieved, it is obviously a good idea to measure charge decay over a range of humidities to assess the humidity-dependence.

Since charge dissipation mechanisms are also the same processes that control the resistivity of the material, the charge decay-time will have the same sensitive dependence on temperature as the resistivity. Other factors that are important are the initial charge or field, the polarity of the charge and the time between recharging the surface when performing consecutive measurements on the same area of sample (Taylor et al, 1987).

## REFERENCES

Blythe AR 1983 "A new conductivity gauge for antistatic films" *Electrostatics'83 IOP Conf Ser No 66* 117-122.

Chubb JN 1988 "Measurement of static charge dissipation" *IOP Short Meetings Ser No14* 73-81.

Chubb JN, Harbour J and Domenichini P 1991 "Instrumentation for on-line measurement of the charge dissipation capabilities of web and coated film materials" *Electrostatics'91 IOP Conf Ser No 118* 153-158.

Klinkenberg A 1958 *"Electrostatics in the Petroleum Industry"* (Amsterdam:Elsevier).

Leonidopoulos G 1991 "Voltage distribution of two thin polymeric film surface resistance measurement systems" *IEEE Trans Instrum Meas* **IM-40** 635-639.

Secker PE 1979 "Monitoring systems for electrostatic powder coating plant" *Electrostatics'79 IOP Conf Ser No 48* 287-293.

Secker PE 1980 "Basic electrostatic measurement techniques" *Oyez Intelligence Report, Electrostatic Hazards* 1-11.

Taillet J 1979 "Method of assessment for the antistatic protection of aircraft" *Electrostatics'79 IOP Conf Ser No 48* 125-133.

Taylor DM and Elias J 1987 "A versatile charge decay meter for assessing antistatic materials" *Electrostatics'87 IOP Conf Ser No 85* 177-181.

Taylor DM, Owen DR and Elias J 1987 "An instrument for measuring static dissipation from materials" *J Electrostatics* **19** 53-64.

Wilson N 1990 "Clothing requirements" *Proc ESD 90 - Electrostatic Damage to Electronics,* ERA report 90-0340.

# CHAPTER EIGHT

# ELECTROSTATIC DISCHARGES (ESD)

Electrical discharges are synonymous with static electricity. Indeed they are often the only outward indication that electrostatic charge is building up to critical levels. They occur when the electric field in the vicinity of a charged object exceeds the breakdown strength of the ambient gas, which is usually air. Electrostatic discharges can be very different in character from one situation to another and depend in a detailed way on the system in which the discharge is initiated. Nevertheless, the basic mechanisms leading to breakdown in any gas are common to all types of discharge.

## 8.1 ELECTRICAL BREAKDOWN IN AIR

Although the behaviour of particular gases under electrical stress may differ in detail, the processes that occur as the breakdown field is approached are similar in all gases. In most industrial situations, though, we are concerned with the breakdown of air. Under uniform electric field conditions and normal temperature and pressure, the electrical failure of air occurs at about 2.7 MV/m, the event being marked by a sudden increase in the conductivity of the gas, the appearance of a luminous spark channel and audible noise. Although the failure is catastrophic it has been possible to study in detail the mechanisms leading to breakdown (Cobine, 1958; Meek and Craggs, 1978). After intensive research over many decades it has been shown that the main features of the breakdown event are well-described in terms of the Townsend $\alpha$- and $\gamma$-processes.

### 8.1.1 Townsend $\alpha$-Process

As a result of cosmic radiation, photoionisation, photoemission and thermal emission processes, there will inevitably be a few adventitiously generated free electrons present in the gap between any pair of electrodes suspended in air.

232

Under the action of an applied field these electrons will drift towards the anode, their velocity controlled by elastic collisions with air molecules. As the field increases so does the average velocity of the electron swarm and, therefore, the average kinetic energy of the electrons.

When the kinetic energy of an individual electron in the swarm exceeds a critical value, then collisions between the electron and a gas molecule become inelastic. The molecule ionises, releasing an electron into the swarm. Since this new electron is indistinguishable from other electrons in the swarm it also can take part in ionising collisions, thereby contributing to the growth in the population of the swarm. What we have described is the Townsend $\alpha$-process which predicts the formation of an electron avalanche. At the onset of avalanching the number of new (or secondary) electrons, $\alpha$, produced per unit path length traversed by the primary electron in the field direction exceeds the number, $\eta$, which attach to gas molecules. As a result, an electron current, $j_0$, emitted from the cathode under uniform field conditions is amplified to an electron current, $j_a$, at the anode where

$$j_a = j_0 \exp(\alpha - \eta)d \qquad (8.1)$$

and d is the distance between anode and cathode. We can see that the electron population grows rapidly once ionisation sets in. The actual growth rate of the electron population is determined by the difference $(\alpha-\eta)$ in the ionisation and attachment coefficients. In highly electronegative gases such as the fluorohydro-carbons (Freons) and $SF_6$, $\eta$ is relatively large and tends to quench the avalanche leading to a higher breakdown strength in these particular gases. In non-attaching gases $\eta$ is small so that avalanches propagate readily and the breakdown strength is generally low.

It should be noted that, for each electron released in an ionising collision, a positively charged ion is also formed. While these ions do not take part in ionising collisions, their presence is important because their associated space charge significantly distorts the electric field in the gas.

Furthermore, for every gas molecule which is ionised, many more only undergo transitions to excited states. When these molecules relax to the ground state they emit photons with wavelengths characteristic of the gas. These photons play a crucial role in the breakdown process as will be discussed below.

## 8.1.2 Townsend $\gamma$-process

The Townsend $\alpha$-process is simply a mechanism which amplifies the electron current already present in the gas. If this initial current is "turned off" then avalanching will also cease. To achieve a self-sustaining discharge a positive feed-back mechanism must be postulated which replenishes the original electrons emitted from the electrode into the gas. Such a mechanism exists.

Some of the photons emitted by excited gas molecules relaxing to the ground state will illuminate the cathode giving rise to photoemission there. A further contribution to this electron emission will occur by positive ion bombardment of the cathode. These secondary processes are known as the Townsend $\gamma$–processes.

In the presence of these processes the electron current arriving at the anode is significantly enhanced, with $j_a$ now being given by

$$j_a = \frac{j_0 \exp{(\alpha - \eta)d}}{1 - \gamma\{\exp{(\alpha - \eta)} - 1\}} \qquad (8.2)$$

where $\gamma$, the Townsend secondary coefficient, is a measure of the secondary electron emission from the cathode. (The theory may be further developed to include (i) the photoionisation of gas molecules by photons released in the primary events and (ii) the space charge effect of positive ions).

The exponential terms in equation (8.2) increase strongly with increasing electric field, and at some critical value, when the denominator goes to zero, an instability in the current sets in and breakdown occurs. Since the secondary processes at the cathode are relatively weak, i.e. $\gamma$ is small, then to a good approximation the instability occurs when

$$\gamma \exp{(\alpha - \eta)d} = 1 \qquad (8.3)$$

Further development of these ideas (Cobine, 1958) leads to the Paschen relation which predicts the complicated way that the sparking potential of a gas depends on gas pressure and electrode spacing, see Figure 8.1.

## 8.2 TYPES OF ELECTRICAL DISCHARGE

Electrical discharges manifest themselves in several different ways depending both on the geometrical arrangement and on the conductivity of the surfaces between which the discharge occurs. Recognising the different types of discharge is important in electrostatics because of the varying degree of hazard that each presents. In the following, a brief description is given of the various discharges that can occur, together with an indication of the industrial situation in which each one is likely to be observed.

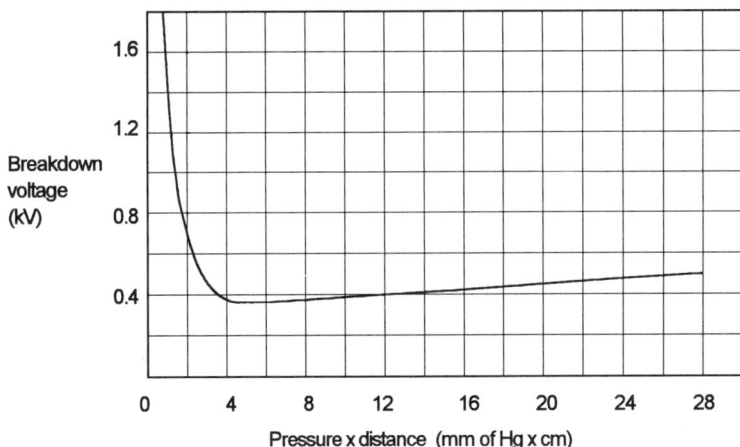

**Figure 8.1** *Breakdown voltage of air at different pressures and with different interelectrode spacings.*

### 8.2.1 Spark Discharge

A spark discharge is the capacitive discharge which occurs between two relatively large metal electrodes when the voltage across them has been raised to the sparking voltage $V_s$. In air at atmospheric pressure the discharge is characterised by a single, highly luminous channel spanning the shortest gap between the conductors. Where the conductors are a pair of parallel electrodes with Bruce or Rogowski profiles (Figure 8.2) to ensure a uniform field in the gap and reduction of edge effects, then the spark can occur anywhere on the electrodes, the exact location being determined by the statistical arrival of adventitious electrons in the gap. The diameter of the spark channel is small because of the constricting effect of the magnetic field associated with the channel current. As the current increases, the magnetic flux which loops around the channel increases in intensity. The effect is to further constrict the channel, thereby increasing both the current density and the gas temperature in the channel. Indeed the latter increases sufficiently to vapourise metal atoms from the cathode, their presence being easily detectable in the optical spectrum of the spark. It is this intense heat of the spark channel that is responsible for initiating the ignition of flammable gases and dust clouds.

In industrial plants, spark discharges can be expected from any conductor which is isolated from earth and which can be charged electrostatically during the processing operation. Examples are the isolated pipe in Figure 2.23 and the isolated container in section 2.7.2(e). Spark discharges from ungrounded humans are also very common in industry and lead to electrostatic problems of varying severity and hazard. (So far as electrostatic phenomena are concerned, humans are no more than thin-film containers filled with electrolyte!!).

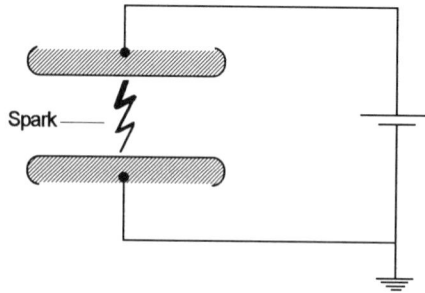

**Figure 8.2**    *Spark breakdown between Bruce profile electrodes.*

### 8.2.2   Brush Discharge

A brush discharge is a type of breakdown which occurs in a relatively long gap in which the electric field is not uniform. An example is the discharge initiated in a gap defined by the sphere-plane electrode geometry of Figure 8.3. When the electric field at the sphere increases above a critical value, the gas in the vicinity of

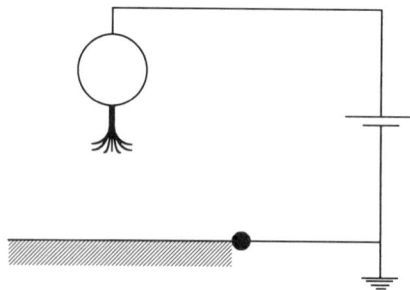

**Figure 8.3**    *Brush discharge from a spherical electrode.*

the sphere suddenly breaks down. A spark discharge is initiated at the sphere but as this spark propagates into the lower field region the spark channel "fans" out into a large number of small, filamentary channels which eventually terminate in the gas. This type of one-electrode discharge is faintly luminous, and is named after its brush-like appearance. The discharge has different characteristics depending on whether the sphere is an anode or a cathode. Once a brush discharge has been initiated, a further increase in the voltage applied across the electrodes can lead to the brush discharge undergoing a transition to a fully developed spark discharge.

It should be noted that in order to initiate a brush discharge the means by which the field is established at the sphere is immaterial. For example, the plane electrode can be replaced by a highly charged plastic film and the sphere may be earthed as in Figure 8.4(a). When the sphere is moved towards such a film, the field at its surface increases until a brush discharge is initiated. Electrons and negative ions generated in the discharge zone migrate to the charged surface, neutralising a region of the film in the immediate vicinity of the sphere (Figure 8.4(b)). Since the charges on the remainder of the film are unable to migrate along the surface of the insulator to the discharge zone, the field at the sphere collapses, the discharge is

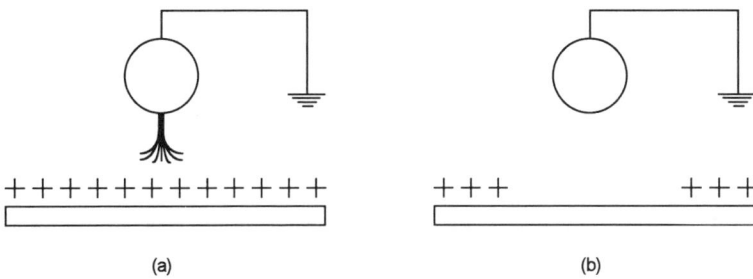

(a)          (b)

**Figure 8.4** *Partial neutralisation of a charged sheet by a brush discharge (a) during and (b) after the discharge.*

quenched and only a fraction of the electrical energy stored in the film is released in the discharge. Moving the sphere parallel to the plane of the film will induce further brush discharges, as each new section of charged film causes the field at the sphere to exceed the critical breakdown value.

A brush discharge to a positively charged surface differs in nature from a discharge to a negatively charged surface. In the former case, negative charges formed in the electron avalanche are accelerated out of the high field region near the electrode, leaving behind a space charge of positive ions, the presence of which decreases the electric field at the charged surface. The formation of localised instabilities known as streamers is thus prevented, so that neutralising negative charges from the discharge deposit in an essentially diffuse circular patch on the insulator. When the surface of the insulator is initially negative, avalanching electrons in the high field near the earthed electrode are accelerated to the electrode. Again a space charge of positive ions forms in the air gap but this time the field at the charged surface is increased and streamer activity is initiated. Neutralising positive charges now approach the insulator along discrete streamer channels which leads to a filamentary (crow's foot) pattern of charge deposition.

The two types of pattern (Lichtenberg patterns) are readily distinguished in the dust that accumulates on plastic sheets. They are particularly obvious on the surfaces of plastic white goods left in storage for some time.

Brush discharges may be observed in a number of industrial situations including the following:

(i)   When a highly charged liquid surface approaches an internal conducting protuberance during tank-filling operations.

(ii)  When any conductive object is lowered into a container of highly charged liquid or powder.

(iii) When conducting objects are inserted into a charged dust cloud or mist.

### 8.2.3   Corona Discharge

The corona, or point,discharge is a special case of a brush discharge in which the radius of curvature of the spherical electrode is small i.e. less than about 1 mm. Consequently, the discharge is located very close to the point. It is a weak discharge exhibiting a faint luminosity which can usually only be seen in the dark. Any item of industrial plant which has a sharp point or edge will go into corona in the presence of a sufficiently high static charge.

Although it is necessary to achieve a critical field at the sharp edge to initiate the corona, once initiated the discharge current remains steady for long periods of time at a value determined by the applied voltage (section 3.4.1).

### 8.2.4   Propagating Brush Discharge

In section 3.2.5 we found that the maximum charge on a freely suspended insulating film was limited by air breakdown to about 50 $\mu C/m^2$ (equation (3.32)). However, when the same film is adjacent to an earthed plate as in Figure 8.5(a) or has a charge of opposite polarity on its back surface as in Figure 8.5(b), the charge density on the upper surface of the film can be significantly greater than the air breakdown limit given above. This is so because the electric field lines from the surface charge are now mainly contained within the film, which has a much greater breakdown strength than the surrounding air. A consequence of this higher charge density is that the electrical energy stored in the film is also greater. If the charge density is sufficiently high in case (a) and an earthed electrode is brought close to the film surface, highly luminous, lightning-like discharges are initiated which emanate radially from the electrode and propagate along the film surface for some considerable distance. The discharge is accompanied by a loud cracking noise.

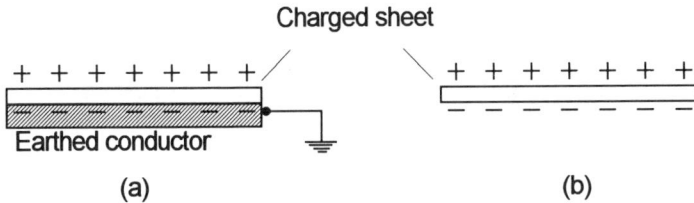

**Figure 8.5** *Situations that can lead to a propagating brush discharge. (a) A charged sheet on an earthed surface and (b) a sheet with equal but opposite polarity charges on its two surfaces.*

The mechanism is readily understood by considering Figure 8.6 which shows the three stages of the discharge. In (a) an earthed sphere is moved towards the highly charged surface of the film. As it does so, the field in the air gap between the surface and the sphere, which is very low initially, increases as field lines switch from the earthed backing plate to the sphere. When the air gap between the sphere and the film surface is comparable with the film thickness, the field at the sphere is sufficiently great to initiate a brush discharge at the sphere (see Figure 8.6(b)). This discharge effectively connects the surface of the film to earth so that a very high electric field is created locally along the film surface. Surface discharges are then initiated which all feed into the original brush discharge as in Figure 8.6(c). The progressive propagation of the high field region along the surface ensures that in such a discharge much of the surface charge originally on the film may be neutralised so that significant quantities of energy are released.

In the case of a free standing film with bipolar charge as in Figure 8.5(b), bringing one earthed electrode to the film surface merely results in discharging very close to the earthing point. However, if both surfaces are earthed simultaneously, propagating brush discharges are once again initiated. Mechanical perforation or a pinhole formed during electrical breakdown of the film can also trigger this type of discharge.

After extensive investigation, Maurer et al (1987) have established that a minimum voltage is required across the film in Figure 8.5(a) if a propagating brush discharge is to be initiated. As can be seen in Figure 8.7, this minimum voltage depends on the film thickness. For films thinner than about 10 µm,propagating brush discharges are unlikely to be initiated.

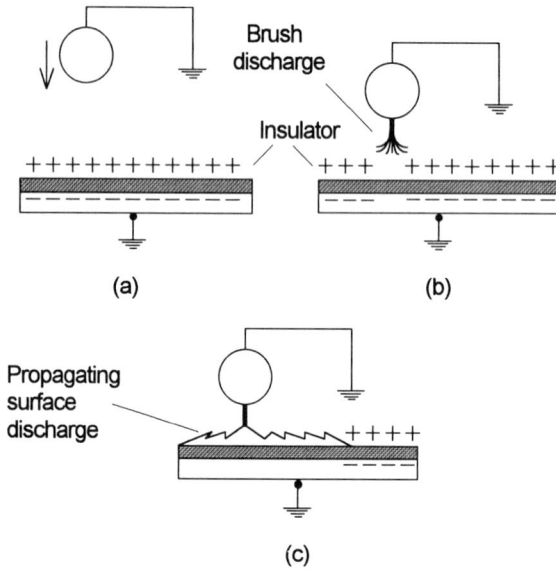

(a)                              (b)

(c)

**Figure 8.6**    *Steps leading to a propagating brush discharge. (a) An earthed conductor approaching a charged sheet resting on an earthed surface. (b) Initiation of a brush discharge which develops into (c) a propagating surface discharge.*

The surface charge densities required for creating the conditions necessary for a propagating brush discharge can often be generated in high speed industrial processes. Possible situations are the pneumatic transport of powders or the high speed pumping of hydrocarbon liquids through insulating pipes (section 3.1.5). Other possible sources of propagating brush discharges are conducting pipes lined with an insulating layer, high speed conveyor belts made from insulating materials and high speed film-winding operations.

### 8.2.5 Cone Discharges

Cone discharges occur during the filling of silos (or other large containers) with relatively coarse granules of an insulating material. They are lightning-like discharges emanating from the wall of the silo and propagating towards the top of the cone formed during the gravity feed of granules into the silo.

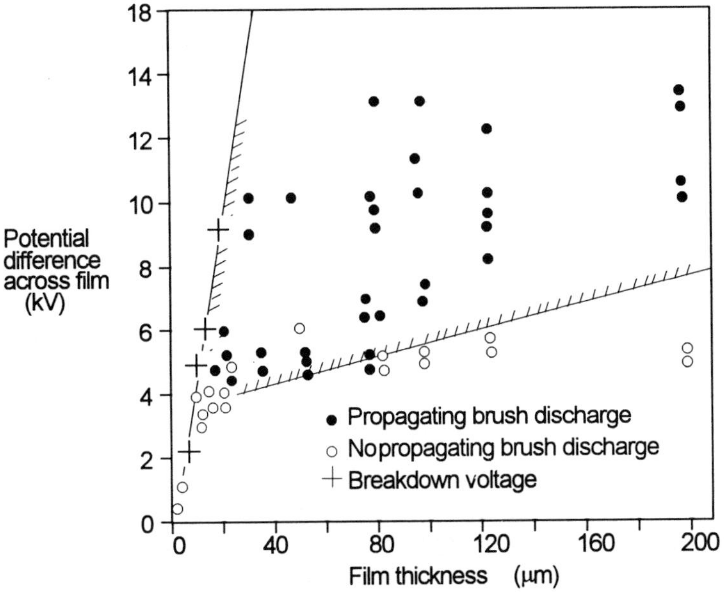

**Figure 8.7**   *Experimentally determined conditions for initiating a propagating brush discharge on polyester and polycarbonate films (Lüttgens and Glor, 1989).*

The origin of this type of discharge is to be found in the concentration of electrostatic charge that occurs during the filling operation. Granules charged elsewhere in the process are compacted during gravity feed into the silo, leading to a sharp increase in volume charge density within the compacted solid. Applying Gauss' theorem to this cylindrical geometry leads us to expect the highest electric fields at the silo wall (section 2.3.4.3), and indeed when granules are fed into centre of the silo, lightning-like discharges are observed to propagate from the wall towards the summit of the heap but are quenched before they reach the low field region in the middle of the silo (Blythe and Reddish,1979; Maurer,1979; Lüttgens and Glor,1989). It should be noted that since these discharges neutralise most of the surface charge on the granules only the topmost layers of newly introduced material are actually involved in the discharges.

When the granule feed-pipe is off-centre, the discharge channels show an asymmetric pattern, with the channels again directed towards the highest point of the cone. This suggests that other factors in addition to the field distribution are important in the initiation and propagation of these discharges, e.g. granules rolling down the heap or the presence of a charged dust cloud above.

## 8.3 SIMULATING SPARK DISCHARGES

We can estimate the explosion or ignition risks associated with certain production processes by subjecting the "product" to a spark discharge of known energy in a controlled laboratory environment. Particularly important in this context is the measurement of the minimum ignition energy (MIE) of a hydrocarbon vapour, powder or vapour/powder mixture. Of increasing importance,in the last decade or so, is the measurement of electrostatic discharge (ESD) damage thresholds in electronic devices,for which a special apparatus based on the human body model has been developed.

### 8.3.1 Minimum Ignition Energy

The ignition energy of a gas, vapour or dispersed solid in air depends strongly on the percentage of flammable material present. At low concentrations the ignition energy is high but decreases to a minimum at some critical concentration before rising again on further increasing the concentration. The lowest energy required to cause ignition of the material, or a mixture of it in critical concentration with air, is known as the minimum ignition energy (MIE). Table 8.1 gives typical values for the MIE of various classes of flammable material.

**Table 8.1**    *Typical range of MIE for various classes of flammable materials (Gibson, 1974).*

| Flammable Material | Typical MIE (mJ) |
|---|---|
| Sensitive detonator explosives | 0.001 - 0.1 |
| Vapour/oxygen mixtures | 0.002 - 0.1 |
| Vapour/air mixtures | 0.1   - 1.0 |
| Chemical dust clouds | 1.0 -5000 |

### *8.3.1.1 Incendivity of Electrostatic Discharges*

The energy released in the various electrostatic discharges discussed in section 8.2 and consequently the incendivity of the discharges differ from one type to another. The following summarises the presently accepted guidelines for evaluating the ignition risk associated with different types of electrostatic discharge.

### (a) Spark Discharge

Spark discharge hazards are associated primarily with isolated metal conductors in a process plant. The energy stored on such a conductor is determined by its capacitance, C, and by its potential, V, at the instant the spark is initiated. As a first approximation, the whole of the stored energy, $\frac{1}{2}CV^2$, is assumed to be dissipated in the spark and if this exceeds the MIE of the surrounding flammable vapour or dust cloud a hazardous situation is created.

We must include in this class of discharge the sparks that occur from electrostatically charged humans. Indeed, discharges from personnel are a common cause of electrostatically induced ignitions in industry and have been shown in the laboratory to be capable of igniting both flammable vapour/air and sensitive dust/air mixtures.

We can consider the capacitance of the human body to be in the range 80 - 300 pF depending on the manner in which the person is insulated from earth. In most cases the capacitance is determined by the thickness of the sole on the footwear being worn by the person. It should be noted that in cases where conducting footwear is being worn, it is still possible to charge the body capacitance so long as the person is standing on an insulated platform or walking over insulating flooring. In such cases, however, the effective capacitance may be less than the normally assumed minimum value of 80 pF.

When there is a spark from a person, the discharge current must flow through the body resistance which is determined in the main by the skin, the remainder of the body volume being essentially a weak electrolyte of reasonable conductivity. During the discharge, some of the stored energy is dissipated in the body resistance and, therefore, will not be available to the spark. Experiments have shown that a consequence of this is that sparks from personnel are only likely to cause an ignition when approximately 2 - 3 times the MIE is stored on the body capacitance. A rule of thumb which has been suggested for the chemical industry is that sparks from personnel should be considered capable of igniting all flammable mixtures with an MIE less than 100 mJ (BS 5958, Control of Undesirable Static Electricity, Part 2).

## (b) Brush Discharge

The energy released in a brush discharge initiated from a conducting sphere brought near a charged insulating surface is found to depend on the size of the sphere and on the area of the charged region on the insulating sheet. Brush discharges from an earthed sphere 20 mm in diameter approaching a charged insulator with an area as small as $20 \times 10^{-4}$ m$^2$ were found to be sufficiently energetic to ignite the most readily ignitable mixture of hydrogen and air. Brush discharges from spherical electrodes 20 mm in diameter should be considered as having an equivalent energy of about 1 mJ (Gibson, 1974; Gibson and Lloyd,

1965). In other words, brush discharges from such an electrode have the igniting power of a spark discharge of energy 1 mJ even though the energy dissipated in the brush discharge may actually be greater. Glor (1981) has shown that, for brush discharges from a spherical electrode with a diameter of 70 mm to a charged surface of area 0.13 m², the equivalent energy of the discharge is likely to be in the range 3 - 4 mJ.

Experiments by Tolson (1981) have shown that brush discharges to negatively charged surfaces are more incendive than those to positively charged surfaces. This finding is not too surprising in view of the differences in the nature of the discharges in the two cases (section 8.2.2).

The relatively low values given earlier for the effective or equivalent energy of a brush discharge should be contrasted with measured values of 5-15 mJ for the total energy dissipated in a brush discharge from a 15 mm diameter electrode to a charged surface $80 \times 10^{-4}$ m² in area. We must conclude, therefore, that the reduced incendivity of brush discharges compared with spark discharges must be related in some way to the different spatial and temporal energy distribution in the two cases.

To date, no-one has been able to demonstrate in the laboratory that brush discharges can ignite pure dust/air mixtures. This is surprising since many dusts have MIE's in the range of a few mJ. It appears that the presence of dust can modify the nature of the discharge. Lüttgens and Glor (1989) suggest that dust settling on the electrode leads to the initiation of weaker brush or corona discharges.

(c) Corona Discharge

The equivalent energy of corona discharges is particularly low so that only very sensitive gas/air mixtures (MIE<0.025 mJ) are capable of being ignited. Therefore, apart from atmospheres containing the most easily ignitable concentrations of acetylene, hydrogen and carbon disulphide, corona discharges may be regarded as non-incendive.

(d) Propagating Brush Discharge

The maximum energy that can be released in a propagating brush discharge can be estimated by considering the charged sheet in Figure 8.5(a) to be equivalent to a parallel-plate capacitor of the same geometry. Thus, knowing the surface charge density, $\sigma$, and film thickness, d, we readily see from equation (2.53) that the maximum energy stored per unit area of the sheet prior to the initiation of the discharge is given by

$$U = \frac{\sigma^2 d}{2\varepsilon\varepsilon_0}. \tag{8.4}$$

For a surface charge density of $2\times10^{-3}$ C/m$^2$ on a sheet 100 $\mu$m thick with a relative permittivity of 2.25 this corresponds to a stored energy of about 10 J/m$^2$. Based on such calculations and a few experimental investigations, Lüttgens and Glor (1989) suggest that propagating brush discharges must be assumed to be capable of igniting all vapour/air and dust/air mixtures with MIE less than 1 J. Hence propagating brush discharges are potentially very dangerous and have a high probability of giving rise to ignition.

### (e) Cone Discharge

Insufficient studies have been carried out to determine the incendivity of cone discharges. Maurer (1984) suggests an equivalent energy of <10 mJ for such events, though it has to be stressed that this value must be treated as a provisional guideline based on calculations and an experimental study within a 100 m$^3$ silo.

The relative incendivity of the various discharges described above is summarised in Figure 8.8 which provides a comparison of the minimum energy of combustible materials and the range of energies that can be liberated in each type of discharge.

### 8.3.1.2 Measuring the MIE of a Gas/Air Mixture

Figure 8.8 shows that the MIE of flammable gases and vapours is well within the range of energies available in electrostatic discharges. It must be assumed, therefore, that all flammable gases and vapours can be ignited by electrostatic discharges. Hence, we should conclude perhaps that measuring the MIE of gases and vapours is somewhat academic in terms of quantifying the electrostatic risk. If, for some other reason, it is necessary to obtain an accurate value for the MIE, then the classic method described by Lewis and Von Elbe (1961) is preferred.

A diagram outlining the main features of their measurement system is shown in Figure 8.9. The ignition chamber is a stainless steel "bomb" internal diameter 12.5 cm. The gas or vapour mixture under test is introduced into the chamber which contains a spark gap comprising two stainless steel electrodes. Across the electrodes are a bank of precision air capacitors which can be switched into the circuit either singly or in combination. The selected capacitor is charged slowly from a high voltage DC supply through a high value resistor which effectively decouples the supply from the electrode system during the actual breakdown period. In BS 5958, Part 1, the recommended value for this resistor is between $10^8$ and $10^9$ $\Omega$,

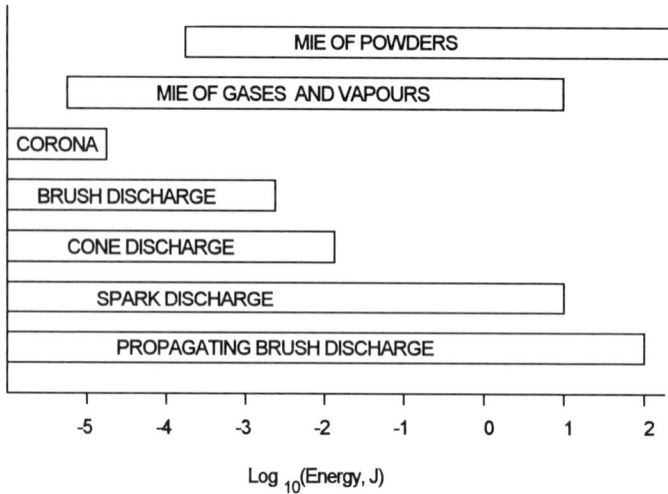

**Figure 8.8**    *Comparison of the energy ranges of some types of discharges with the MIE of flammable and combustible materials (Lü ttgens and Glor, 1989).*

though Lewis and Von Elbe report using a bakelite rod with a resistance of about $10^{11}$ $\Omega$. When the voltage across the capacitor exceeds the breakdown voltage of the interelectrode gap a spark discharge is initiated. Since the RC time constant of the capacitor and series resistance is much greater than the spark duration, only the

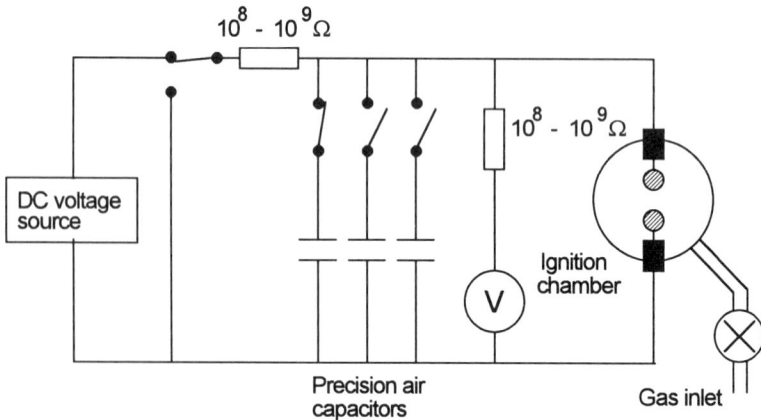

**Figure 8.9**    *Arrangement used by Lewis and Von Elbe (1961) to measure the MIE of gases and vapours.*

energy stored on the capacitor is dissipated in the spark. After the discharge, the capacitor automatically recharges ready for the next spark. The sparking voltage is measured with an electrostatic voltmeter connected to the high voltage electrode through a decoupling resistor whose magnitude is again in the range $10^8$ - $10^9$ $\Omega$.

The test begins with the lowest value of capacitance connected across the electrodes. If ignition does not occur, the spark energy is gradually increased by connecting capacitances of increasing value until ignition does occur. The MIE is taken to be the lowest spark energy that ignites the most easily ignitable concentration of the gas or vapour.

Careful experimentation has established that the measured MIE depends on the electrode size. Large electrodes absorb heat from the spark causing the discharge to quench, particularly at small interelectrode gaps. Therefore, small electrodes should be used for the measurement. However, for diameters less than 2 mm corona emission just prior to sparking can complicate the measurements. BS 5958 Part 1 recommends an electrode diameter of 10 mm with an electrode gap not less than 2 mm.

Introducing a resistance between the capacitor and the gap increases the time duration of the spark which can also affect the measured MIE. Rose and Priede (1958) found that,as long as the resistance was < 100 $\Omega$, the ignitability of an air/methane mixture was independent of the resistance. As the resistance increased, so the MIE increased slowly owing to damping of the oscillatory nature of the discharge. Above about 1 k$\Omega$, which corresponded with critical damping of the circuit, the MIE decreased slowly to a minimum which was recorded with about 10 k$\Omega$ in the circuit.

Replacing the series resistance with an inductor so as to enhance the oscillatory nature of the spark has been shown to increase the ignitability of a gas (Kono et al, 1976).

When measuring the MIE of very sensitive mixtures the capacitance used in the tests may need to be as small as 1 pF. For accurate results it is important to include the capacitance of the electrodes and connecting leads when computing the total energy stored.

### 8.3.1.3 Measuring the MIE of a Dust/Air Mixture

Measuring the MIE of dust clouds is identical in principle to that described above for gases. However, the measurement is complicated by the presence of many more experimental parameters, not the least of which is the generation of a uniform dust cloud of reproducible concentration. In the case of gaseous mixtures, sparks can be initiated repeatedly under virtually identical conditions. For dust clouds it is necessary not only to regenerate the cloud each time but also to trigger the spark at exactly the same time in each experiment. The triggering of the spark may also be influenced significantly by space charge fields associated with static charge on the powder itself.

Despite such problems, in the last few years an internationally accepted test procedure has been developed. This has shown that earlier results obtained in a test described by the US Bureau of Mines, which ignored the transformer losses in the test circuit, may overestimate the MIE of some powders by as much as a factor of 2 to 5. As a result the electrostatic risk during the handling of these particular powders may have been seriously underestimated.

It is suggested now that the dust to be tested is dispersed in an explosion vessel such as "the Hartmann tube" or "the 20-litre laboratory sphere" and exposed to a spark discharge from a capacitor of known value. Figure 8.10 shows a section through a typical Hartmann tube. Approximately 0.5 to 1 g of the powder to be tested is placed in the dispersion cup. Since the MIE of fine dusts is much lower than for coarse dusts, only that fraction of the powder passing through a 200 mesh grid,i.e. particles less than 75 μm,should be tested.

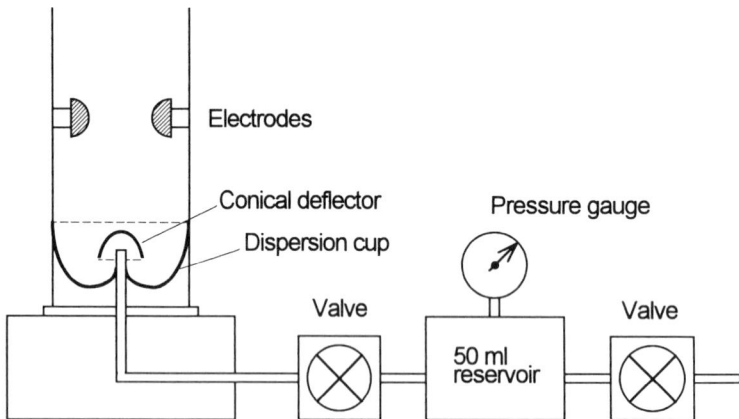

**Figure 8.10**    *Hartmann tube for measuring the MIE of powders.*

The sample is dispersed by a controlled jet of compressed air from a 50 ml reservoir pressurised to 8 - 10 bar. The air jet is deflected by the conical deflector into the dispersion cup where it picks up the powder sample and disperses it throughout the tube forming a cloud of suspended particles. At the appropriate moment a spark is triggered and the experiment repeated with gradually increasing spark energy until ignition occurs.

It is recommended that initiation of the spark is delayed for as long as possible after powder dispersion in order to achieve ignitions under the least turbulent conditions possible. It is further suggested that samples are dried for 24 hours at 50°C under atmospheric pressure before the measurement is carried out since the presence of moisture in the sample can increase the MIE.

The spark itself may be generated using the charging circuit in Figure 8.9 which allows repeated sparking in the dust cloud. By continually redispersing the powder a range of dust concentrations will then be tested.

In an alternative test method which allows the spark to be triggered at a fixed time following the powder dispersion, a third electrode may be used to initiate breakdown of the main gap. In such an arrangement (Figure 8.11) the capacitor across the main electrodes is charged to just below the breakdown voltage. After dispersing the powder, the voltage of the auxilliary electrode is raised until it sparks.This low-energy auxilliary spark then triggers the breakdown of the main gap. Obviously, since the MIE measurement is based on the energy dissipated in the main spark, care must be taken to ensure that the energy of the auxilliary spark is small by comparison.

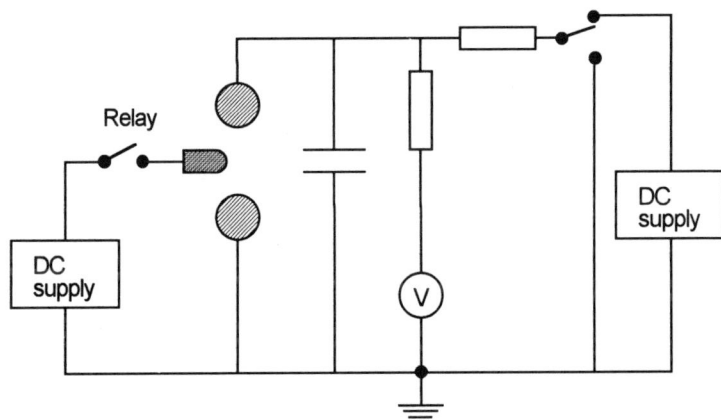

**Figure 8.11**    *3-Electrode system for triggering a spark discharge at a preset time.*

A third method that has been suggested is the moving electrode technique. Here the interelectrode gap is set at a sufficiently large distance that even with a high voltage across the capacitor the electric field in the gap is well below that required for breakdown. As soon as the powder is dispersed, one of the electrodes is moved quickly towards the other until a spark is initiated. The energy of the spark is again assumed to be equal to the energy stored on the capacitor.

BS 5958 (Part 1) recommends the "trickle charging" method while Glor (1989) and Rogers (1991) give all three possibilities.

The MIE of a dust cloud has been found to depend on a number of electrical parameters associated with the test circuit. Therefore, in the interests of standard-isation, no matter which method is chosen the spark generator must satisfy the following important criteria (Glor, 1989):

(a) The circuit inductance must be greater than 1 mH, though this is only important for dusts with high MIE. If required, the additional inductance may be included between the low voltage electrode and earth.

(b) The series resistance between the charged capacitor and the electrodes must be $< 5\,\Omega$.

(c) Electrode diameter equal to 2 mm with an electrode gap of at least 6 mm (cf. 10 mm diameter electrodes and a gap greater than 2 mm in BS 5958, Part 1). Prior to testing, it should be confirmed that the electrodes do not produce corona before sparking.

(d) Low-inductance, surge-proof capacitors.

(e) Low-capacitance electrode arrangement.

(f) The electrodes must be well insulated from the remainder of the apparatus to ensure that the test voltage is readily attainable without adventitious discharges occurring elsewhere in the test system.

### 8.3.2 Human Body Model

The sensitivity of electronic devices and systems to electrostatic discharges has made it imperative to devise a test method that will realistically simulate the effects of discharges on such devices and systems. Charged personnel were perceived to be the most likely source of electrical discharges in these cases, with the spark energy being released either directly to devices during handling or to the casing of operational systems. Consequently, the circuit specified in IEC 801-2 (1st edition) for testing electronic devices and systems was designed to simulate the behaviour of the human body. The system (Figure 8.12) consists of a simple charging circuit in which the body capacitance is represented by the 150 pF capacitor and the body resistance by the 150 $\Omega$ resistor. The test voltage is variable from 2 kV to 15 kV so as to cover the range of potentials likely to be encountered on humans. The dimensions of the high-voltage discharge probe have been chosen to simulate an outstretched finger. A current return strap is connected between the low voltage terminal of the capacitor and the earthed reference plane below the equipment under test.

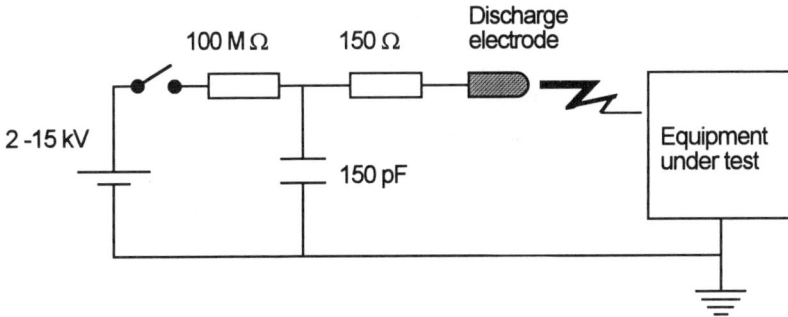

**Figure 8.12**    *Human body model (IEC 801-2, 1st edition).*

In use, the capacitor is charged to the test voltage with the probe tip held away from the equipment under test. The probe is then moved towards the equipment until a discharge occurs. The recommended waveform for such a circuit is shown in Figure 8.13. It should be noted that if the surface of the equipment is insulating no discharge will occur.

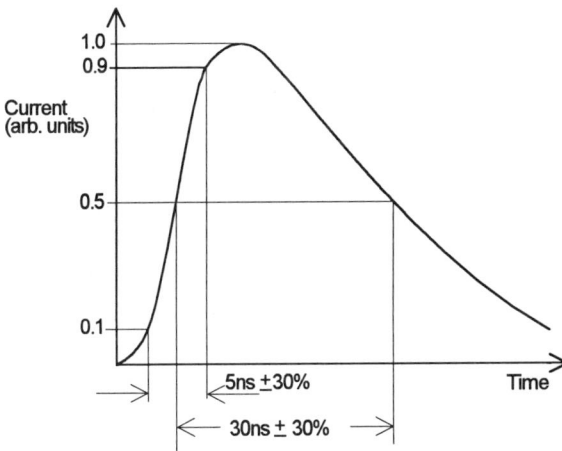

**Figure 8.13**    *Time-dependence of a spark from the human body model (Jones, 1990).*

Although the human body model has served the very important purpose of standardising the test procedure, there is still much discussion concerning the details of the test circuit. Of immediate concern is that discharges from humans to earthed surfaces do not usually display the "smooth" characteristic shown in Figure 8.13. The rise-time is found to depend both on the speed of approach of the tip to the equipment under test and on the test voltage (Jones, 1990).

Measurements with high speed oscilloscopes have shown that the waveform associated with sparks from humans are more complex than those from the human body model and are more closely simulated as a sharp current peak lasting about 1 ns superimposed on the waveform resulting from the human-body model circuit.

Accordingly, a new test circuit is under discussion (IEC 801-2, 2nd edition (draft)) in which it is recommended that tests are carried out with the probe in contact with the equipment under test and a relay is used to connect the previously charged capacitor to the equipment as in Figure 8.14. This new test certainly reproduces the rapidly rising initial transient but the waveform is strongly oscillatory and more complex than discharges from humans.

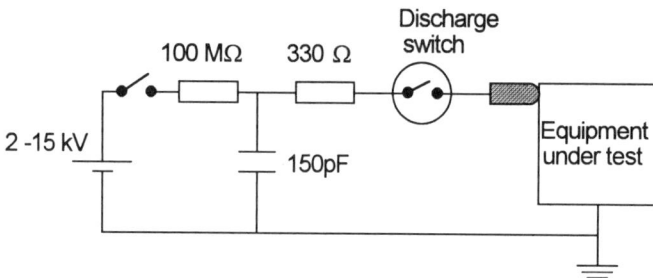

**Figure 8.14**    *Proposed new human body model (IEC 801-2, 2nd edition - draft).*

The test as described above is suitable for quantifying the sensitivity of complete systems to electrostatic discharges. A similar circuit has also been used in a commercially-available ESD test equipment for determining the sensitivity of integrated circuit chips to electrostatic discharges (Shaw and Enoch, 1983). The device under test is located in a special holder and earthed through one of its pins. Subsequently, each pin of the device is connected in turn to the previously charged capacitor. The voltage on the capacitor is gradually increased and each pin subjected to discharges of increasing energy until the device is damaged.

Such tests have proved invaluable for demonstrating the effectiveness of protection circuits and for showing the importance of circuit layout in increasing the immunity of devices to electrostatic discharges (Enoch et al, 1983).

## 8.4 MONITORING SPARK DISCHARGES

### 8.4.1 Radio Frequency Detection

The intense magnetic field associated with the spark discharge current causes severe constriction of the spark channel. The rapid changes in both the electric and magnetic fields in the spark gap resulting from spontaneous contractions and expansions of the spark channel give rise to the emission of electromagnetic radiation in the form of radio waves. AM (amplitude modulated) radio techniques have been used to detect the presence of hazardous spark discharges from their associated rf emission (Butterworth,1979; Chubb,1975; Lüttgens and Glor,1989).

Although radio detection has been used to detect sparks for some time (Chubb et al, 1973), no agreement has been reached on the optimum frequency for detecting these events. Chubb (1975) concentrated on the 10 to 100 MHz frequency range while Lüttgens and Glor (1989) suggest the intermediate frequency range 450 - 480 kHz, which avoids interference from normal broadcasting frequencies. For the range 150 kHz to 3 MHz a receiver such as that based on the ZN414Z shown in Figure 8.15 is suitable for detection. The demodulated signal can be used to trigger an acoustic or optical alarm or can be plotted on a recorder. Chubb (1975) used the detected signal to trigger a photographic camera inside the hold of an oil tanker during washing and was able to correlate the occurrence of sparks with a particular orientation of the rotating, high-pressure water hose used for cleaning the hold.

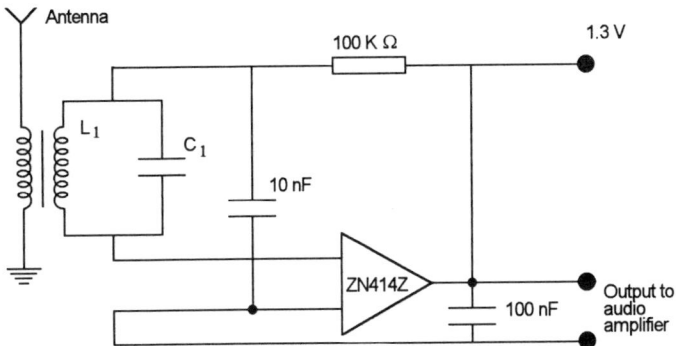

**Figure 8.15**   *Tuned-circuit receiver for the rf detection of spark discharges. $L_1$ and $C_1$ are chosen to give reception at a particular frequency.*

Most industrial environments are electrically very noisy so that many adventitious signals totally unrelated to electrostatic sparking events may also be picked up by the probe circuitry. To identify unambiguously the occurrence of a spark, a second detector should be located outside the vessel or chamber in which sparking is expected and/or an additional detector at other places within the vessel. Co-incident and anti-coincident circuits can then be used to eliminate the extraneous signals.

The sensitivity of radio techniques is such that sparks with energies lower even than the MIE of gas mixtures can be detected. Therefore, if sparks are detected these need not necessarily be incendive. A corollary to this, of course, is that if no sparks are detected the process is operating safely.

In principle at least, the rf emission spectrum of a spark should provide inform-ation on the geometry of the sparking body and the potential of the body at spark-over. This was demonstrated by Butterworth (1979) using the basic circuit shown in Figure 8.16. A discharging body was suspended above a ground plane in a reflectionless environment and charged through a $10^{10} \ \Omega$ resistor formed from a 2m length of 2 mm bore plastic tubing filled with isopropanol. The signals were detected with a short monopole antenna located on the earth plane some 8 m away. By limiting the length of the antenna to $\ll \lambda/4$ at the highest frequencies, where $\lambda$ is the wavelength, the response is only a weak function of frequency. The antenna was connected via a short length of 50 $\Omega$ coaxial cable to the input of a Tektronix 7L12 spectrum analyser positioned below the earth plane to avoid spurious reflections and other interference.

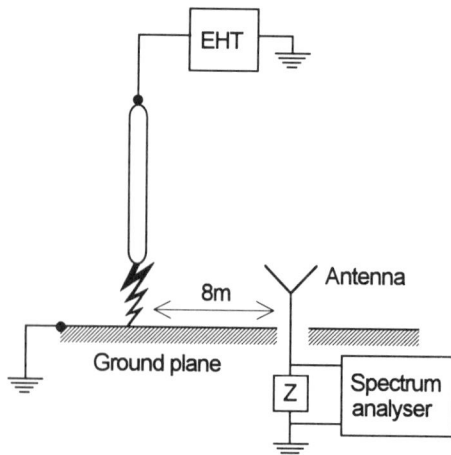

**Figure 8.16**    *Arrangement used by Butterworth (1979) to investigate the effect of the geometry of the sparking body on the characteristics of the spark.*

Discharges from spherical bodies were essentially broad-band with no features relating to geometry. At low breakdown potentials, frequency components up to at least 2 GHz were present. As the breakdown potential increased, the highest frequencies progressively decreased and were limited to about 500 MHz at 15 kV.

Discharges from cylindrical bodies with a base machined to form a hemisphere gave particularly interesting results. The emission spectrum now contained strong bands at frequencies satisfying the relation

$$L = (2n - 1)\frac{\lambda}{4} \qquad (8.5)$$

where L is the length of the cylinder and n an integer.

Butterworth also investigated the discharges to charged bodies from long metal poles (3.5 and 6.5 m long) projecting from the ground plane. Once again he detected well-defined bands in the emission spectrum which satisfied equation (8.5) but this time smaller peaks were also distinguished corresponding to half-wave modes, i.e. satisfying the relation

$$L = n\frac{\lambda}{2}. \qquad (8.6)$$

Radio frequency emission from spark discharges could, therefore, provide much useful information concerning the body from which the discharge is initiated. Unfortunately, many hazardous electrostatic sparks occur inside closed metallic enclosures so that radiowaves from the spark may undergo multiple reflections before arriving at the detector. The signal then becomes dependent on the location of both the source and the detector, with resonant modes of the enclosure being superimposed on, and generally swamping, the emission spectrum of the spark.

Despite these difficulties, radio techniques can be used to discriminate between spark and corona discharges. Not only is corona emission considerably weaker, but in addition the emission spectrum is not detectable beyond about 200 MHz (Butterworth, 1979).

Finally, it should be noted that radio techniques are often used for detecting lightning storms. The long length of lightning strokes means that they emit strongly at about 10 kHz. With suitable detectors, lightning strikes may be detected up to 1000 km away, and by combining radio detection with measurements of the atmospheric electric field it is possible to follow the path of electrical storms with some accuracy.

### 8.4.2 Measuring the Current Flow in Electrostatic Discharges

Measuring the current flowing in an electrostatic discharge requires consider-able care if meaningful results are to be obtained. In a purely capacitive circuit, Figure 8.17(a), the circuit current, i, is given by

$$i = \frac{dq}{dt} = C\frac{\partial v}{\partial t} + v\frac{\partial C}{\partial t} \qquad (8.7)$$

where q and v are the instantaneous values of charge and voltage on a capacitor of magnitude C. Thus a current will flow in response to (i) a change in the voltage across the capacitor and (ii) a change in the capacitance of a charged capacitor, e.g. by moving the capacitor plates closer together.

This current, known as the displacement current, may be measured by placing a small resistor, R, in series with one of the capacitor plates as in Figure 8.17(b) and monitoring the voltage drop across it. For rapidly changing signals R must be very much smaller than the reactance, $1/2\pi fC$, of the capacitor where f is the signal frequency.

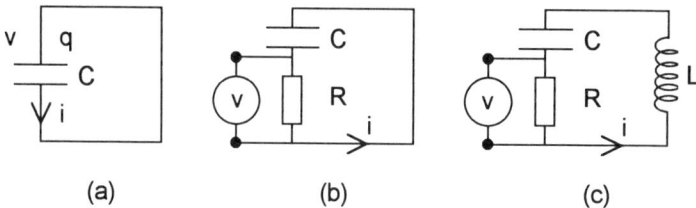

(a)                    (b)                    (c)

**Figure 8.17**    *To measure the short-circuit current in (a) a small resistor must be placed in series with the capacitor as in (b). The effect of lead inductance (c) is to cause current oscillations in the circuit as shown in Figure 8.18.*

The leads connecting the various circuit components will also have an associat-ed inductance, so that a more correct representation of the measuring circuit is that shown in Figure 8.17(c), where L represents the stray inductance.

The influence of this inductance becomes apparent at high frequency, i.e. for short periods of time after initiating a transient current flow in the circuit. For example, when $2\sqrt{L/C} > R$, then the rapid discharge of capacitor C will give rise to current oscillations in the circuit. When R is small, the frequency of the oscillation is nearly equal to the resonant frequency $f_0 = 1/2\pi\sqrt{LC}$ of the circuit. As energy flows back and forth between the capacitor and the inductor, some is dissipated in the resistor, damping the current oscillations as shown in Figure 8.18.

The ratio $(i_{n+1}/i_n)$ of successive current maxima is readily shown (Noakes, 1956) to be equal to $\exp(-R/2f_oL)$.

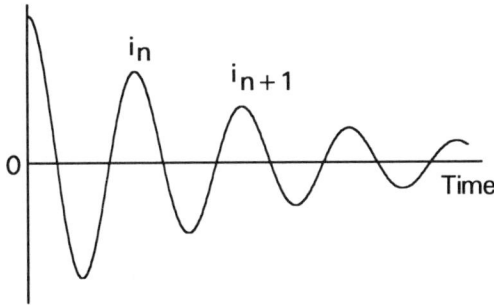

**Figure 8.18** *Current transient during discharge of capacitor in Figure 8.17(c).*

When mobile charges are present in the gap between the plates of the capacitor, for example when a spark occurs, an extra current flows through the measuring resistor. This extra current depends both on the number and on the velocity of the charges in the gap.Consider a charge Q moving a distance dx in a direction normal to a pair of parallel-plate electrodes separated by a distance d as in Figure 8.19. It is easily shown (Shockley, 1938) that this results in the transfer of a charge $\delta q = Q(\delta x/d)$ from one electrode to another through the external circuit. If the charge Q moves with velocity $u = \delta x/\delta t$ from one plate to another, then the current flowing in the external circuit is given by

$$i_Q = \frac{Q}{d}\frac{\delta x}{\delta t} = Q\frac{u}{d} \qquad (8.8)$$

It is the magnitude and duration of this "real" current that is of importance in the ignition of flammable materials. We should contrast this with the human body model for testing electronic devices and systems, where both the real and displacement currents will be important; both contribute to the current flow in devices and both contribute to electromagnetic interference.

We might assume that measuring the current in a spark would be a trivial matter once the problem of circuit resonances has been dealt with. This may not necessarily be the case as can be seen by considering a spark discharge from an isolated but charged conducting sphere to a large, conducting ground plane as in Figure 8.20(a). At first sight, it seems reasonable to measure the current by connecting the conducting plane to earth through a small resistor (of negligible inductance to avoid ringing) and measuring the voltage across it with a high speed oscilloscope. Somewhat surprisingly perhaps, this method will not give the true discharge current waveform.

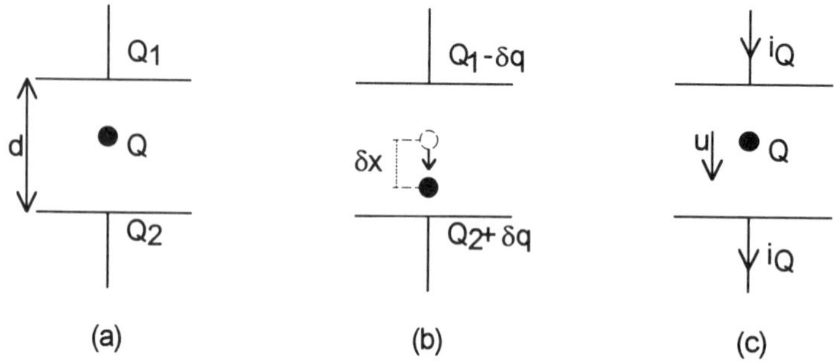

**Figure 8.19** *Charges induced on electrodes by a charge Q (a) initially and (b) after moving a distance δx. (c) When Q moves at constant velocity a displacement current flows in the external circuit.*

To understand why, we need to remember that as the positively charged conductor in Figure 8.20(a) approaches the conducting plane it induces negative charges on the surface of the conducting plane. The flow of these charges from ground gives rise to a current through the measuring resistor even before a spark is initiated. If, immediately prior to the discharge, the positive charges on the sphere have induced an equal number of negative charges on the surface of the plate, then the ground plane is acting as a Faraday pail. (Even though the "pail" does not totally surround the charge, all the flux from the charged sphere terminates on the plane,which is exactly the requirement in a Faraday pail).

When the spark is initiated the current through R will consist of two equal and opposite components. The first will arise from the flow of real positive charges in the gap,i.e. the spark current, i, that we are trying to measure. The second will be the opposing displacement current, $i_D$, arising from the collapse of the voltage, v, across the sphere-plane capacitance. The displacement current is given by

$$i_D = C\frac{dv}{dt} = C\left(\frac{1}{C}\frac{dq}{dt}\right) = \frac{dq}{dt} \tag{8.9}$$

where q is the instantaneous charge on the sphere. The rate at which this charge decreases is equal to i, the rate at which charge is conducted away in the spark. Therefore, if the initial charge on the sphere was $q_0$ the charge at some time t after the initiation of the spark is

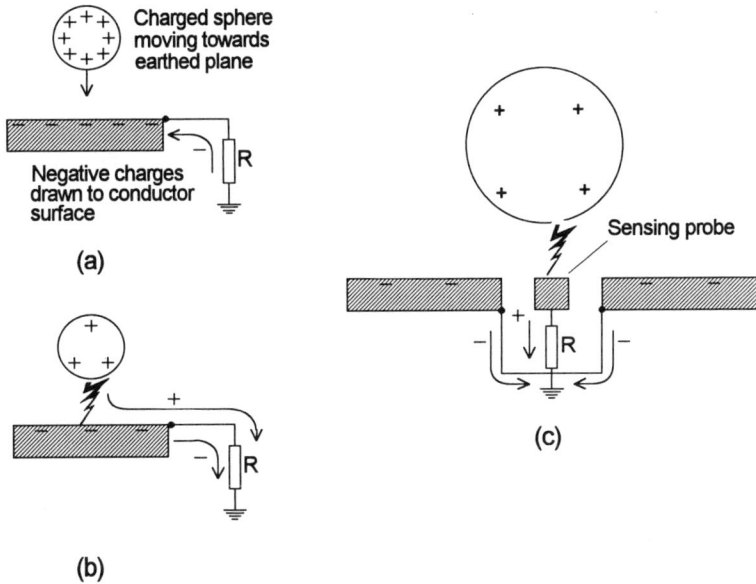

**Figure 8.20**    *(a) An approaching positive charge induces a negative charge in the earthed plate. (b) When sparking occurs both the positive and induced negative charges flow to earth through R. (c) Positive charges flowing to earth through the sensing probe are separated from the negative charge flow.*

$$q = q_0 - \int_0^t i.dt \qquad (8.10)$$

which upon comparing with equation (8.9) shows that $i_D$ must equal - i.

Thus, in the special case where just prior to the spark all the flux lines from the charged sphere are coupled to the plate, the real and displacement currents exactly balance so that the net current flowing through the measuring resistor during the spark is zero. In other words, for every positive charge that flows to ground from the sphere, an induced negative charge must also flow to ground as in Figure 8.20(b).

To make an accurate measurement of the spark current, it is necessary to connect the measuring resistance R to a small probe located in the centre of, but insulated from, the earthed plane as shown in Figure 8.20(c). Now, so long as the spark terminates on the probe, the spark current may be separated from the displacement current, since the majority of the charges initially induced on the ground plane flow directly to ground, by-passing the measuring resistor and thus do not contribute to the signal across it.

From their studies Makin and Lees (1981) concluded that the greatest accuracy should be obtained when the probe diameter is less than 0.025 mm, though Chubb and Butterworth (1982) argue that a larger probe can be used successfully when a correctly designed measuring impedance is used. To minimise inductive effects, the latter authors used a resistance of 1 Ω formed from ten 10 Ω, $\frac{1}{8}$W solid carbon resistors connected in parallel and located very close to the probe.

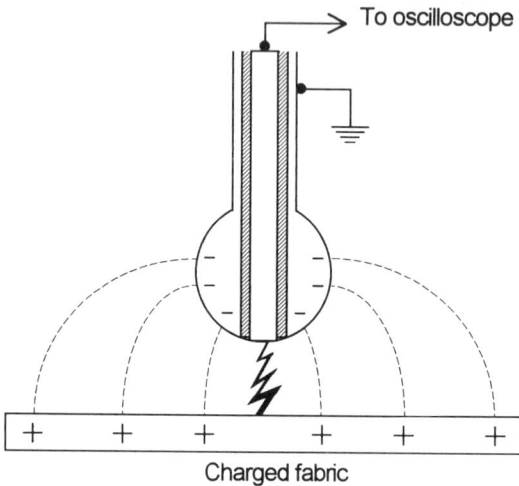

**Figure 8.21**    *Guarded probe for measuring the discharge current from charged fabric (Wilson, 1985).*

The measurement principle outlined above can be used to measure the current flowing in sparks from insulating surfaces. Wilson (1985) used the arrangement in Figure 8.21 to investigate the incendiary behaviour of spark discharges from textile surfaces. The probe consists of a stainless steel sphere surrounding but insulated from an internal cylindrical core. The 3 cm diameter sphere forms an earthed shield on which most of the induced charges appears when charged fabric is brought to the probe. The spark from the fabric terminates on the cylindrical electrode which is connected to a storage oscilloscope to measure the spark current.

Of course the approach described above is only necessary in situations such as that in Figure 8.20 where the spark current is supplied mainly from charges stored on the capacitance between the sphere and the earthed plane. In many instances this is not the case. In the human body model in Figure 8.12, for example, the spark current is almost entirely supplied by charges stored on the capacitor connected across the spark gap. Prior to the discharge, these charges are not coupled to the earthed plane, so that the displacement current will be small compared with the true spark current. Thus, the simple circuit in Figure 8.20(b) will give accurate results so long as the sphere-plane capacitance is small compared with any additional capacitances connected across the gap.

Finally, it should be noted that the currents flowing during a spark discharge are sufficiently great that they may be detected inductively by means of a small sensing coil wound between the low-voltage electrode and earth as shown in Figure 8.22. Using this arrangement Davies (1990) has recorded the waveforms of spark discharges from a person charged to 3 kV with a time resolution of a few nanoseconds. It is assumed here, of course, that the person-to-sphere capacitance is small compared with the capacitance between the person and ground.

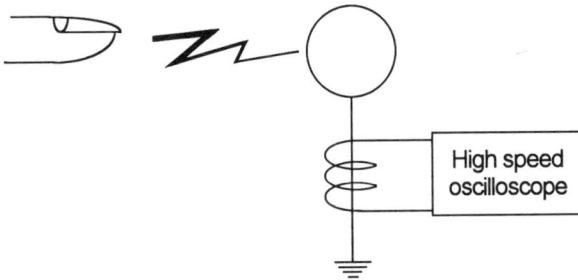

**Figure 8.22**   *Simple circuit for measuring spark current from charged humans.*

## 8.5 CONCLUDING REMARKS

Over the last two decades or so, considerable progress has been made in understanding the nature of the risk presented by electrostatic discharges in industry. Methods for handling vapours and gases are well-documented and the extent of the risks is understood. The situation for powders is still evolving.

## REFERENCES

Blythe AR and Reddish W 1979 "Charging on powders and bulking effects" *IOP Conf Ser No48* 107-114.

Butterworth GJ 1979 "The detection and characterisation of electrostatic sparks by radio methods" *IOP Conf Ser No 48* 97-105.

Chubb JN 1975 "Practical and computer assessments of ignition hazards during tank washing and during wave action in part-ballasted OBO cargo tanks" *J Electrostatics* **1** 61-70.

Chubb JN and Butterworth GJ 1982 "Charge transfer and current flow measurements in electrostatic discharges" *J Electrostatics* **13** 209-214.

Chubb JN, Erents SK and Pollard IE 1973 "Radio detection of low energy electrostatic sparks " *Nature* **245** 206-207.

Cobine JD 1958 *"Gaseous Conductors: Theory and Engineering Applications"* (New York: Dover).

Davies DK 1990 "Auditing the workplace" *Proc ESD 90 - Electrostatic Damage to Electronics* ERA Report 90-0340.

Enoch RD, Shaw RN and Taylor RG 1983 "ESD sensitivity of NMOS LSI circuits and their failure characteristics" *Proc EOS/ESD Symp* **EOS-5** 185-197.

Gibson N 1974 "Safety problems associated with electrostatically charged solids" *Dechema Monograph* **72** 343-355.

Gibson N and Lloyd FC 1965 "Incendivity of discharges from electrostatically charged plastics" *Br J Appl Phys* **16** 1619-31.

Glor M 1981 "Ignition of gas/air mixtures by discharges between electrostatically charged plastic surfaces and metallic electrodes" *J Electrostatics* **10** 327-332.

Jones B 1990 "Human-body model test - IEC 801-2" *Proc ESD 90 - Electrostatic Damage to Electronics* ERA Report 90-0340.

Kono M, Kamagai S and Sakai T 1976 "Ignition of gases by two successive sparks with reference to frequency effect of capacitance sparks" *Combustion and Flame* **27** 85-98.

Lewis B and von Elbe G 1961 *"Combustion, Flames and Explosions of Gases"* (London: Academic Press).

Lüttgens G and Glor M 1989 *"Understanding and Controlling Static Electricity"* ( Ehningen bei Boblingen: Expert-Verlag).

Makin B and Lees P 1981 "Measurement of charge transfer in electrostatic discharge" *J Electrostatics* **10** 333-339.

Maurer B 1979 "Discharges resulting from electrostatic charging in large storage silos" *Chem Ing Tech* **51** 98-103.

Maurer B 1984 *VDI-Berichte* **494** 119. (Cited in Lüttgens G and Glor M 1989 *"Understanding and Controlling Static Electricity"* ( Ehningen bei Boblingen: Expert-Verlag)).

Maurer B, Glor M, Lüttgens G and Post L 1987 "Hazards associated with propagating brush discharges on flexible intermediate bulk containers, compounds and coated materials" *IOP Conf Ser No 85* 217-222.

Meek JM and Craggs JD (eds) 1978 *"Electrical Breakdown of Gases"* (Chichester: Wiley).

Noakes GR 1956 *"Electrical Fundamentals"* (London: HMSO).

Rogers RL 1991 "Sensitivity of flammable gases, vapours and powders to ignition by electrostatic discharges" *Seminar Documentation E7746: Electrostatic Hazards in Industry* (London:IBC Technical Services Ltd).

Rose HE and Priede T 1958 "Ignition phenomena in halogen-air mixtures" *Proc. 7th Symp. on Combustion,* (London: Butterworth) 436-445.

Shaw RN and Enoch RD 1983 "A programmable equipment for electrostatic discharge testing to human body models" *Proc EOS/ESD Symp* **EOS-5** 48-55.

Shockley W 1938 "Currents to conductors induced by a moving point charge" *J Appl Phys* **9** 635-636.

Tolson P 1981 "Assessing the safety of electically powered static eliminators for use in flammable atmospheres" *J Electrostatics* **11** 57-69.

Wilson N 1985 "The nature and incendiary behaviour of spark discharges from textile surfaces" *J Electrostatics* **16** 231-245.

# AUTHOR INDEX

Abdullah, M. 175,178
Akutsu, K. 7
Anzai, H. 94

Bäckstrom, G. 153,199
Bamford, W.D. 155
Bassett, J.D. 153,188
Baum, E.A. 199
Bauser, H. 91,94
Berta, I. 42
Bhuiyan, L.B. 68
Bloodworth, G.G. 133
Blum, L. 68
Blythe, A.R. 133,143,230,241
Böcker, H. 162
Bomenichini, P. 230
Bowdler, G.W. 162,173,175,178
Boyd, H.A. 182
Bradshaw, E. 166
Bright, A.W. 133,182,204
Brooks, H.B. 161,162
Brown, A. 95
Bruce, F.M. 163,165,182
Brussel, M.K. 39,48
Burt, J.P.H. 11
Butterworth, G.J. 253,254,255,260

Carlson, C. 4
Cassidy, E.C. 156
Chalmers, J.A. 139,203
Chapman, D. 68
Chowdry, A. 91
Chubb, J.N. 146,147,150,151,155,204,
  227,230,253,260
Clark, H.E. 4
Cobine, J.D. 102,114,115,232,234
Cones, H.N. 156
Corbett, R.P. 153,182,188
Coste, J. 96

Craggs, J.D. 64,232
Crane, J.S. 11
Crofts, D.W. 81
Cross, J.A. 11,12,48,96,192
Das Gupta, D.K. 122
Davies, D.K. 89,90,95,199,261
de Myer, G. 74,76
de Smet, M. 74,76
Defandorf, F.M. 161
Denbow, N. 204
Dessauer, J.H. 4
Diserens, N.J. 32
Doughty, K. 122
Duke, C.B. 94,95

Ebert, H. 203
Elias, J. 225,228,230,231
Elsdon, R. 93
Endo, S. 94
Enoch, R.D. 3,252
Erents, S.K. 253

Fabish, T.J. 94,95
Faraday, M. 84
Finar, I.L. 95
Forrest, R.H. 151
Fujibayashi, K. 7

Gallagher, T.J. 64,73
Gallo, C.F. 115
Gavis, J. 76,78
Gerhard, R. 207
Gerhard-Multhaupt, R. 199
Gibson. H.W. 94
Gibson, N. 76,81,242,243
Gilbert, W. 84
Glor, M. 239,241,244,245,250,253
Gouy, G. 68
Grahame, D.C. 68

Haenan, H.T.M. 96
Harbour, J. 230
Harper, W.R. 84,85,87,96
Hasse, H. 202
Hays, D.A. 94
Henderson, D. 68
Hersh, S.P. 97
Higham, J.B. 155
Hines, R.L. 108
Horvath, T. 42
Huang, Y. 11
Hughes, J.F. 6,182,189
Hughes, K.A. 199
Husain, S.A. 166

Inaba, H. 7
Ishida, K. 7

Jones, B. 251,252

Kamagai, S. 247
Kelvin, Lord 160
Kesavamurty, N. 166
Klinkenberg, A. 68,74,223
Klopffer, W. 91
Kono, M. 247
Koszman, I. 76,78
Krupp, H. 91
Kuffell, E. 175,178,179,182

Lamo, W.L. 115
Lees, P. 260
Leonidopoulos, G. 221
Lewiner, J. 207
Lewis, B. 245,246
Lewis, T.J. 75,78,81,91,94,199
Lloyd, F.C. 76,243
Lowell, J. 85,91,93,95,96
Lüttgens, G. 239,241,244,245,253

Macken, W.J. 139
Mapleson, W.W. 139,149
Masuda, S. 7,8,9
Matsumoto, Y. 7
Maurer, B. 239,241,245

McAllister, D. 32
Makin, B. 260
Meek, J.M. 64,232
Menon, K.B. 166
Misuno, A. 7
Mitchell, F.R.G. 93
Montgomery, D.J. 96,97
Moore, A.D. 12
Moreau-Hanot, M. 153
Morisseau, D. 207
Moyle, B.D. 189
Nayfeh, M.N. 39,48
Nickell 104
Noakes, G.R. 257
Nordhage, F. 153,199

Odam, G.A.M. 151
Ohara, K. 96
Olivier, J.P. 71,75
Outhwaite, C.W. 68
Owen, D.R. 225,230,231
O'Dwyer, J.J. 64

Patey, G.N. 68
Pauthenier, M.M. 153
Pechery, P. 96
Peek, F.W. 112
Pethig, R. 11,13,124
Petry, W. 199
Pohl, H.A. 11
Pollard, I.E. 150,151,253
Post, L. 239
Priede, T. 247

Rabenhorst, H. 91
Reddish, W. 133,143,241
Rogers, R.L. 250
Rose, H.E. 247
Rose-Innes, A.C. 85,91,93
Rutgers, A.J. 74,76

Safford, F.J. 167,168
Sakai, T. 247
Schaffert, R.M. 4,123,124
Schon 77

Secker, P.E.  104,143,145,146,149,169,
  199,213,214,219
Sennet, P.  71,75
Sessler, G.M.  125,207
Shaw, R.N.  3,252
Shinohara, I.  94
Shockley, W.  257
Silsbee, F.B.  161
Smith, J.  133
Smith. J.R.  32
Smythe, W.R.  39,48,50,117
Stern, O.  68
Sweet, R.G.  6
Sze, S.M.  88,89

Taillet, J.  213
Taylor, D.M.  75,78,79,80,81,91,104,225,
  228,230,231
Taylor, R.G.  3,252
Tedford, D.J.  182
Tolson, P.  244
Toomer, R.  199
Torrie, G.M.  68
Townsend, J.S.  114
Trump, J.G.  167,168

Unger, B.A.  3,97

Valleau, J.P.  68
Van de Graaff, R.J.  167,168

Van der Meer, D.  2,82
Van der Minne, J.L.  68,74
Van der Weerd, J.M.  183,208
Von Elbe, G.  245,246
Von Seggern, H.  207
Vos, B.  204
Vosteen, R.E.  153,170
Vyverberg, R.G.  113,124

Walmsley, H.L.  77,79,80
Wang, X-B.  11
Warburg, E.  115
Washizu, M.  7
Waters, R.T.  149
Waterton, F.W.  165
West, J.E.  125,207
Westgate, C.R.  91
Whitlock, W.S.  139,149
Williams, T.P.T.  75,78,81
Wilson, A.D.  208
Wilson, N.  226,260
Woodford, G.  77,80
Wunsch, D.C.  156

Yamamoto, F.  94
Yaratich, M.  133

Zichy, E.L.  199
Zimmer, W.A.  96

# SUBJECT INDEX

Acceptor states 89,93,95
Acetylene 244
Additives for liquid dielectrics 62,77,81
Alternating voltage dividers 175
Anthracene 94
Anthraquinone 94
Antistatic footwear 211-
Aperiodic damping 165
Applications of static electricity 4-
Arrhenius temperature dependence of resistivity 220,222
Atomisation 6,107
Atoms 17
Attracted-plate voltmeter 161-

Back-discharging 83,98,100
Back-ionisation 103-,121-,155,189
Back-tunneling 83,93,98-
Ballistic probe 153-
Ballooning effects 81
Band gap 88
Bar codes 6
Beneficiation 10-,129
Bias current of amplifier 131-
Biological cells 11
Biotechnology 11
Bipolar particle charging 120
Boltzmann's equation 70
Bonding to ground 211
Boxer charger 7-
Breakdown of air 1,21,23,100,102,232
Breakdown strength 64
Bruce ellipsoid voltmeter 165-
Bruce profile electrodes 163,168,181,235
Brush discharge 82,100,101,236-,243-
Bulking effects 241

Capacitance 16,43-,175
- addition 52-

- standard 50
Capacitance of
- coaxial cylinders 46
- concentric spheres 49
- cylinder/plane configuration 48
- human body 52,250-
- non-concentric spheres 50-
- parallel cylinders 47
- sphere/plane configuration 50-
Capacitor 60-,65
Carbon resistors 260
Carbon disulphide 244
Cell separation 11
Characterisation of materials by
- charge decay 223-
- by resistivity 215
Charge 56,61,185-,257
- accumulation 63
- amplifier 130
- carriers 59,64,68
- clouds 2
- density 23,70,72
- distribution throughout a volume 201-
- flow to a filter 204
- probe for pneumatic systems 192
- separation 67-
- transport 4,12
Charge decay 61,63,210-,223-
- from films and laminates 224-
- in liquids 223-
Charged dust cloud 241
Charged particle probe 155
Charged surfaces 56,193-,237
Charge-decay time-constant 230
Charge-to-mass ratio 10,186
Charging
- by filters 82
- limits 100-,107.118
- of dusts 7

267

- of liquids 68-
- of filaments 80
- of personnel 109-,235,243,261
- of transforemer oil 80
Charging time constant 118-
Cleanliness 16,65
Conduction band 87-
Conductive footwear 110
Conductivity 59,78,88,90
Cone discharge 240-,245
Contact charging of solids 82-,188
Contact potential difference 17
Corona 21,102-,110-,130,171,182
- charging 4,6,7,10,67,112,116,122,
  188,227-
- current/voltage characteristic 114
- current density 115-
- discharge 100,238,244
- ions 124,155
- onset 102,149
- poling 122
- probe 208
- source 214-
- threshold voltage 112-,122
Corotron 4,123-
Cosmic radiation 125,232
Coulomb's law 18-
Crop-spraying 109
Current 56,64,73,114-
- flow in electrostatic discharge 256-
Cylinder of charge 24-

Dependent parameters 55
Dielectric
- boundaries 33-,38-
- constant 18,56
- interfaces 34-
- strength 64
Dielectrics 36,56
Dielectrophoresis 11
Dielectrophoretic cell sorting 11
Diffusion
- charging 121
- coefficient 71
Direct voltage dividers 171

Discrete energy states 85
Displacement 21,24,34
- current 256
Dissociation of contaminants 68
Donor states 89,95
Double-layer 68-
- thickness 72,74-
Drift zone 111
Dust 2,18,23,65,130,192,242,243,244,
  247-

Ebert tube 203
Einstein relation 71
Electrets 125
Electric charges 16-,23
Electric curtain powder collection 6-
Electric
- field 16-,19,21,23,26,129-,255
- field at interfaces 34
- field lines 63
- field strength 20
- shock 110
Electrochemical potential difference 17
Electromagnetic interference 257
Electron
- affinity 88
- attachment 179,233
- avalanche 111,233
- beams 125
- charge 17
- tunnelling 86,90
Electronic charge, 17,70
Electronic device failure 3,250
Electrons 17,59,84
Electrophotography 4-,103,112,129
Electrostatic
- attraction 1,42
- cling 1,18,36,42,130
- containment 12
- discharges (ESD) 3,232-
- force effects 159
- generators 12
- painting 6,108,214
- precipitation 7,10,112,118,121,155
- pressure 42,108

- separation 10-
- theory 16-
Electrostatically-induced fires and
  explosions 2
Electrostatic/gravitational effectiveness
  ratio 186
Electro-optical effects 155-
Electro-osmotic flow 68
Emulsification 82
Energy
- gap of a semiconductor 88
- released in a spark 242-
- stored in a capacitor 53-,58
Equipotentials 30
Equivalent circuit 61,64,210,213-
ESD sensitivity 3,12,14,55,242

Faraday pail 73,104,186-,205,258
Fermi energy 85-
Field measurement errors 132-
Field meter 193,205,208
Field mill 104,136-,195,201,203,204,215,
  224,227
- applications 148-
- voltmeter 167-
Field
- penetration through aperture 143
- polarity detection 139-
- reduction by compartmentalisation 26
Filamentary discharge channel 236
Film winding electrostatic effects 1,3,104
Flammable
- atmospheres 150,151,211
- gases 242,245
- materials 12,242
- vapours 242,245
Flashover voltages for coaxial sphere
  electrodes 179-
Fluidised beds 12
Flux density 20-,24,34
Forces between charges 17-
Footwear testing 211-
Free ions 188
Frictional charging of solids 82-

Gamma-rays 125
Gauss' theorem 23,24-,36,46,49,102,210,
  241
Generating voltmeter 167-
Grounding of metal equipment 3
Guard electrode 131,160,161,173,196,
  216,219

Hammett constant 94-
Hartmann tube 248
Hazards 103
High voltage terminals 171,174
High-stability carbon film resistors 172
Human body model 250-
Humidity 3,16,65,84,103,211,222-,231
Humidity effect on
- breakdown voltage 179
- charge-decay rate 231
- minimum ignition energy (MIE) 249
- surface resistivity 222
Hydrogen 244
H.V. capacitor 176

Ignition 65,130,242
Image charges 36-
Image force 6,38,40,42
Immiscible liquid effects 82
Impulse voltage measurement 177
Independent parameters 55
Induced charge detector 188
Induction
- charging 10,35-,67,104-,130-,187
- probe 130-,155,195,196-,228
Ink-jet printing 6,106
Insulating
- film 205-
- liquids 12
- pipe effects 49,81
- rollers 3
- web 145-,193
Insulators 12,16,56,60,65,83-,89,205,
  210-,220
Integrated circuits 3,252
Interleaving vane voltmeter 166
Intrinsically safe field mill 150

Ion transfer 96
Ion wind 155
Ionisation zone 111
Ionising collisions 111
Ions 17,59,84
Irradiation of spark gap 179
Isopropanol 254

Kelvin attracted disc electrometer 160
Kerr cell 156

Laplace equation 32,116
Laser printer 5
Lichtenberg patterns 1,98,238
Lightning strikes 255
Limiting surface charge density 100-
Lines of force 19
Liquids 23
Localised band gap energy states 89

Magnetic tape 2
Man-made fibre 56
Material transfer on contact 84
Measurement of MIE 245
Metal film resistors 173
Method of Images 40-
Millikan oil drop experiment 155
Mineral beneficiation 10,129
Minimum ignition energy (MIE)
   2,55,242-
Mixing 82
Mobility 12,59,88,220
Molecules 17,58
Monitoring spark discharges 253-
MOS technology 3

Nebulisers 109
Negative ions 70,73,111,124
Neutral energy 91
Non-accessible conductors 212-
Non-contact voltage measurement 169
Null-field detection technique 153

On-line surface-charge-decay gauge 228-
Organic liquids 56,64,68

Organic vapours 65,129
Oscillation-type voltmeter 165-
Outlet effects on streaming current 79
Ozone 82,94

Paschen relation 234
Pauthenier limit 118,154
Permittivity of free space 18
Photoconductor 4
Photoelectrons 125
Photons 233,234
Piezoelectric
- sensor 12
- transducer 106,122
Pinhole formation 122
Pipeline flow 68,73-
Plastic film capacitors 2
Pneumatic transfer charging 191
Pockel cell 156
Point/plane electrode geometry
   110,115,155
Poisson's equation 31-,70-,114
Polarisation 11,35,57-,105,115,122,220,
   222,230
Polyethylene 88, 94
Polymer electrification during extrusion
   81
Polymeric materials 56,64
Polystyrene 94
Polyvinyl alcohol 95
Positive ions 17,70,73,111,124,237,238
Potential 16,28-,55-,148
Potential dividers 170-
Powder
- clouds 23
- coating 6,103,155,214
- handling 104
Pressure pulse measurement of charge in a
   solid 206-
Pressurised gas insulation
   162,165,167,176
Process speed 2
Propagating brush discharge 238,244
Protection
- against ESD 3,12

- circuits for buffer amplifiers 175,183, 219-
Proton 17
Pyroelectric transducer 122

Radio frequency spark detection 253-
Radio interference 151
Radioactive source for non-contact electrical coupling 182,183
Radioactivity 17,125
Rayleigh limit for a stable droplet 108
Relative humidity - see humidity
Relative permittivity 18,45,56-
Relaxation time 60,62-,71,78,90,106,154, 198,205,210,223
Resistance 59,65,172,210-
Resistivity of powders 218-
Response time 61
Rogowski profile electrodes 181,235
Rubbing effects 96

Scaling up 2
Scorotron 123
Sensing plate
- induction probe/field mill 130-
Separation 10-
Settling 82
Shock to personnel 2
Shockley states 94
Signal-to-noise ratio 199
Solid carbon resistors 260
Space charge measurement 203
Space potential 183-
Spark 23,100,235-,237,242-
- channel 64,235
- energy 2,242-
- gap 65,179-
Specific charge 102,104
Sphere of charge 25
Spray electrification 82
Standards for electrostatics 12-,150,179, 210,211,225,226,243,245,247,248,250, 251, 252
Static
- neutralisers 3,21,104,110,124

- dissipating footwear 152
- sensitive electronic components 211, 212,250
Step-function voltage measurement 177
Streamers 237,238
Streaming current 73-,81,205
- in transformers 80
Surface
- charge density 193-
- charging 17,69-
- resistivity 60,63,221-
- states 91-

Tamm states 94
Temperature effect on
- charge-decay rate 231
- surface resistivity 222
- volume resistivity 220
Textile fibres 1
Textiles 1,260
Time constant 61
Toluene 76
Townsend
- α-process 111,232-
- γ-process 232,233
Transformer oil
-electrification of 80-
Traps 12,205,230
Triboelectric
- series 96,224
- spray guns 103
Triboelectrification 6,10,82-,96-,204,213, 225

Use of light-pipes at high voltage 169
UV light 17,125

Vacuum level 85
Valence band 87-
Valence electrons 85
Van de Graaff generator 1,167
Very Large Crude Carrier (VLCC) explosions 2
Vibrating probe
- field meter 152-

- probe system  199
Virtual earth amplifier  131-,187,191,219
Voltage  29,55-,61,65,159-
- dividers  170-
Volume charge density  30,241
- in a gas  201-
- in a liquid  204
- in a solid  205
Volume resistivity  59,63,90,215-
- of liquids  217-
- of solids  216-

Washing of fuel-cargo tanks
 2,82,204,253
Waterfalls  82
Web charging  101

Wien bridge oscillator  164
Wimshurst machine  1
Wire-wound resistors of nickel/copper
 alloy 173
Work function  85

Yarn  18,36,41,81

Xerography  4
X-ray film  2
X-rays  125

Zener-diode safety barriers  150
Zero drift of amplifier  131-
Zeta potential  75,86